Social and Gender Analysis in Natural Resource Management

Social and Gender Analysis in Natural Resource Management

Learning Studies and Lessons from Asia

Edited by
RONNIE VERNOOY

China Agriculture Press

SAGE Publications
New Delhi • Thousand Oaks • London

Copyright © International Development Research Centre, Canada, 2006

All rights reserved. No part of this book may be reproduced or utilized in any form or by any means, electronic or mechanical, including photocopying, recording or by any information storage or retrieval system, without permission in writing from the publisher.

The boundaries and names shown on the maps in this publication do not imply official endorsement or acceptance by the International Development Research Centre.

Jointly published in 2006 by

Sage Publications India Pvt Ltd
B-42, Panchsheel Enclave
New Delhi 110 017
www.indiasage.com

Sage Publications Inc
2455 Teller Road
Thousand Oaks, California 91320

Sage Publications Ltd
1 Oliver's Yard, 55 City Road
London EC1Y 1SP

and

International Development Research Centre
PO Box 8500, Ottawa, ON, Canada K1G 3H9, info@idrc.ca / www.idrc.ca

China Agriculture Press
18 Maizidian Street, Chaoyang District, Beijing 100026, China, www.ccap.org.cn

Published by Tejeshwar Singh for Sage Publications India Pvt Ltd, phototypeset in 10.5/12.5 pt Minion by Star Compugraphics Private Limited, Delhi and printed at Chaman Enterprises, New Delhi.

Library of Congress Cataloging-in-Publication Data

Social and gender analysis in natural resource management: learning studies and lessons from Asia/edited by Ronnie Vernooy.
    p. cm.
    Includes bibliographical references and index.
    1. Rural development—Asia—Case studies. 2. Women in development—Asia—Case studies. 3. Natural resources—Management—Research. I. Vernooy, Ronnie, 1963–

HC412.5.S63	333.708'095—dc22	2006	2005033467

ISBN: 0–7619–3462–6 (Hb)   81–7829–612–8 (India-Hb)
         0–7619–3463–4 (Pb)   81–7829–613–6 (India-Pb)
         1-55250-218-X (IDRC e-book)

Sage Production Team: Payal Dhar, Ashok R. Chandran, Rajib Chatterjee and Santosh Rawat

Contents

List of Tables 7
List of Figures 9
List of Boxes 10
Preface by Ronnie Vernooy 11
Acknowledgements 13

1. Integrating social and gender analysis into natural resource management research 17
   Ronnie Vernooy and Liz Fajber

2. The social and gendered nature of ginger production and commercialization: A case study of the Rai, Lepcha and Brahmin-Chhetri in Sikkim and Kalimpong, West Bengal, India 37
   Chanda Gurung and Nawraj Gurung

3. Strengthening market linkages for women vegetable vendors: Experiences from Kohima, Nagaland, India 65
   Vengota Nakro and Chozhule Kikhi

4. Enhancing farmers' marketing capacity and strengthening the local seed system: Action research for the conservation and use of agrobiodiversity in Bara District, Nepal 99
   Deepa Singh, Anil Subedi and Pitamber Shrestha

5. Empowering women farmers and strengthening the local seed system: Action research in Guangxi, China 129
   Yiching Song and Linxiu Zhang with Ronnie Vernooy

6 Social and Gender Analysis in Natural Resource Management

6. Creating opportunities for change: Strengthening the social capital of women and the poor in upland communities in Hue, Viet Nam   155
   Hoang Thi Sen and Le Van An

7. Herder women speak out: Towards more equitable co-management of grasslands and other natural resources in Mongolia   181
   H. Ykhanbai, Ts. Odgerel, E. Bulgan and B. Naranchimeg

8. Similarities and differences: From improved understanding to social transformations   207
   Ronnie Vernooy and Linxiu Zhang

9. Social and gender analysis is essential, not optional: Enhanced capacities and remaining challenges   225
   Ronnie Vernooy and Linxiu Zhang

About the Editor and Contributors   237
Index   243

List of Tables

1.1	Key features of women in development (WID) and gender and development (GAD) approaches to research	29
1.2	Key features of the six studies	31
1.3	Methods, tools and skills used in the six case studies	32
2.1	Demographics of the two study regions compared with national data, 2001	41
2.2	Size and ethnic composition of households in the selected villages	48
3.1	An overview of vegetable vending enterprises in Kohima	75
3.2	Designated market niches for vegetables vendors provided by the Kohima Town Committee	78
3.3	Number of vegetable species sold each month by women vendors from Pholami	80
3.4	Costs and benefits of a trip to the Kohima market for vendor 1 (INR)	85
3.5	Costs and benefits of a trip to the Kohima market for vendor 2 (INR)	86
3.6	Comparison of retail and wholesale prices per unit of vegetables and fruits obtained by a vendor from Pholami village (INR)	86
3.7	Options for increasing and improving production of marketable vegetables and fruits in Merema and Tsiese Basa	93
4.1	Sources of seeds across wealth categories	116
4.2	Involvement of men and women in seed production across wealth categories	120

5.1	Changes in out-migration in Anhui, Qinghai and Guangxi, 1997–2003	132
5.2	Basic information about the five research sites	138
5.3	Key characteristics of maize production at the research sites	139
5.4	Men and women's perceptions of who manages household resources and activities	140
5.5	Men and women's perceptions of who makes decisions about the management of resources and activities	141
5.6	Comparison of variety selection criteria between women and men farmers in Guangxi villages	143
6.1	Land use in Hong Ha and Huong Nguyen communes	162
6.2	Population of Hong Ha and Huong Nguyen communes	162
6.3	Wealth ranking of commune households	163
6.4	Access to training by wealth category in the two communes	166
6.5	Number of women and poor households involved in interest groups	174
7.1	Participation of men and women in the protection and restoration of natural resources according to community members	194
7.2	Pasture rotation plan designed by the Arjargalant community women's group	196
7.3	Changes in income structure of herder households, 2001–03 ($n = 36$)	198

List of Figures

1.1	Theory of action	28
2.1	Location of the study sites in the Sikkim region	40
3.1	The Nagaland region showing the location of the research sites	68
4.1	Map of Nepal showing the research site	102
4.2	Reasons for growing modern varieties of rice	114
4.3	Reasons for growing landraces of rice	114
4.4	Period of seed replacement for three socioeconomic categories of farmers	115
4.5	Existing seed marketing channels for Kachorwa farmers	117
4.6	Rates for seed exchange across socioeconomic categories	118
4.7	Seed selection methods according to wealth category	119
4.8	Labour used at various stages of seed production by socioeconomic category	121
4.9	Participation by men and women in decision making and selling (rich farmers only)	122
4.10	New seed marketing channels	124
5.1	Trends in out-migration in Anhui, Qinghai and Guanxi by sex	133
5.2	Location of the research sites in southern China	137
6.1	The Viet Nam research site	161
6.2	Access of men and women to training courses	166
6.3	Source of decisions about training topics	168
7.1	Location of the study sites	186
7.2	The participation of men and women in farming and household work ($n = 84$)	193

List of Boxes

3.1	The NEPED project	71
3.2	*Gajo-jotho*: A promising wild vegetable for domestication	81
3.3	Fencing: A leap towards sedentary agriculture	82
3.4	Insect infestation: A selling point!	83
3.5	Substituting vegetables for rice in TRCs	89
3.6	Creating an overnight storeroom in Kohima	90
3.7	Situation for an average part-time vendor in Tsiese Basa	91
3.8	An alterative marketing chain: The vendors of Pfutsero	97
5.1	The seed fair in Guozhai	149
6.1	People's voices	176

Preface

This is a book about encounters—literally and figuratively speaking. In one way or another, the nine chapters are about encounters among the authors and collaborators and their meetings with rural women and men, and with women and men working for local and national governments and for non-governmental organizations (NGOs), in a number of South Asian and South-East Asian countries. Some of the authors and collaborators are themselves government or NGO employees. The three synthesis chapters and six case studies are also about attempts to create interfaces between the natural sciences and the social sciences, between more descriptive social science research and more transformative participatory research, and between locally situated research 'projects' and wider socioeconomic and sociopolitical structures and processes.

What all contributors have in common is an interest in a better understanding of the societies in which we work and live. In particular, we are concerned about the everyday social differences and inequalities that exist, and how they are changed over time (the process of social differentiation). To varying degrees and in different ways, we also share a desire to look for and explore opportunities for social transformation. Based on our very diverse everyday practices and experiences—encompassing more than the joint research project that this book is about—we have come to understand that this is easier said than done. Becoming involved in social transformation means engaging with politics and power or knowledge struggles; almost by definition, it implies dealing with setbacks and challenges.

The six case studies forming the 'Learning Studies' project describe and reflect on a diversity of efforts to integrate social and gender analysis into natural resource management research. They point to the importance of 'local' history and context, and to the increasingly interlocking 'local' and 'supra-local' forces. In addition, the six cases allow a comparative analysis and the discovery of a number of commonalities.

The cases recognize the steps that researchers are already taking in implementing social and gender analysis research, including questions of class, caste and ethnicity in natural resource management. They all represent 'learning stories'—attempts to deepen our understanding and strengthen our practice.

Through cross-regional exchanges, the six research teams and collaborators supported and encouraged each other to learn along the way, trying to be both reflexive about practice and critical about theories and concepts. The selected cases illustrate 'on-the-ground' examples of challenges and opportunities, successes and disappointments in integrating social and gender analysis. They also highlight a number of methods used and adapted in the very diverse contexts of the Asian region.

The studies make a start at reflecting on what has been done and is being done in organizations in terms of capacity development for the integration of social and gender analysis. They also look at *how* this has been done and is being done, and the enabling and constraining factors that are affecting the process. In addition, they ask how best to support these capacity development efforts in the future.

The authors and contributors share the assessment that the series of encounters enabled by this initiative over a period of more than three years has made a difference. As the chapters will tell, they have indeed deepened our sociological knowledge and strengthened our (action) research skills. They also allowed the building of friendships.

Ronnie Vernooy

Acknowledgements

The research documented in this book and the book itself came to fruition thanks to the support of many people. Women and men farmers and herders shared their ideas, points of view, dreams, concerns and worries with the research teams while teaching us about their way of life in India, Nepal, China, Viet Nam and Mongolia. Extension agents, other researchers and government officials collaborated with the teams, joining the learning caravan on the road to better understanding and action. We acknowledge their contributions to the research, their patience, hospitality and good company.

Special thanks go to John Graham, former IDRC programme officer, now happily enjoying a quieter life. John provided spirited input and critical feedback from start to finish—in the field, during workshops and through long-distance correspondence. He patiently proofread all chapters, highlighting gaps, questionable observations and shaky use of the English language. We have done our best to take his suggestions into account, and he is not to blame for any remaining errors.

Support and encouragement were provided by many other IDRC colleagues from the Sustainable Use of Biodiversity and the Community-based Natural Resource Management teams, the Gender Unit, the communications group, and staff in the Singapore, New Delhi and Ottawa offices. Thank you!

We say *xie xie* to the staff of the Center for Chinese Agricultural Policy (CCAP) in Beijing for their support and good humour, and for demonstrating their singing talents to us fearlessly and skilfully. We acknowledge the perceptive comments of Peng Guangqian when we first dared to venture into a comparative analysis of the six case studies.

Our English writing skills were greatly improved by Sandra Garland. The artful publishing was in the hands of Tejeshwar Singh, managing director of Sage Publications, India, Ke Wenwu, director and senior editor of China Agricultural Press, and Bill Carman, managing editor at IDRC. We also acknowledge the support of Omita Goyal, consultant commissioning editor.

The Sikkim/West Bengal team notes that their region has been long overlooked by most international development organizations. IDRC has been one of the few that has taken an interest in this region and its people, thereby, bringing it into focus on the international development scene. They thank IDRC for this. The process of conducting this study required numerous visits to the villages and long hours interviewing and questioning the farmers. The team would like to express their gratitude to the farmers, who throughout all of this showed immense patience and enthusiasm—and showered the team with hospitality. They are also grateful to the six field assistants for their contribution and invaluable help. Liz Fajber (IDRC) for her guidance and advice, deserves special mention. Nawraj would also like to acknowledge the Indo-Swiss Project, Sikkim, for allowing him to participate in this study.

The Nagaland study authors thank IDRC officials, Ronnie Vernooy, Liz Fajber and John Graham, for allowing them to undertake social and gender analysis research in Nagaland and for supporting them throughout. They also acknowledge the contributions of their peer groups during workshops, discussions and interactions in identifying gaps and shortcomings, improving learning, and enhancing skills and confidence. Through this research, the Nagaland Empowerment of People through Economic Development (NEPED) project has begun to venture into gender issues at other institutions. Thanks also go to Alemtemshi Jamir, IAS, NEPED team leader, for his continuous support, to the Project Operation Unit colleagues, and to other stakeholders including women vendors from four villages, the Kohima Town Committee, the agriculture department, the horticulture department and the district administration. Last but not the least, they acknowledge the contribution made by their family members in terms of moral support, travelling with them at times to research sites, bearing with them when they worked long hours and attached themselves to the computer.

The LI-BIRD authors thank IDRC for financial and technical support throughout the research. They express sincere gratitude to the farming communities of Kachorwa for their constant cooperation. They would like to acknowledge Karna Chaudhary and Phul Kumari

Chaudhary for their continuing support and help during the fieldwork; without it, the study would not have been possible.

The Chinese team gives their sincere thanks to the Guangxi-based social and gender analysis team and to all the farmers from the two seed villages for their great contribution and commitment to the research. They also appreciate the support received from IDRC and from CCAP to carry out the work. Specifically, they thank Ronnie Vernooy, Liz Fajber, Qunying Pan, Bailing Huang and Chengfang Liu for their support throughout the project period. They also thank the other five case study teams for sharing their experiences during the learning process.

The Viet Nam team expresses deep gratitude to all the individuals and institutions who helped in the research. Special thanks go to IDRC for its financial support, guidance and advice. They also thank the Hue University of Agriculture and Forestry for its encouragement and administrative assistance. The research could not have been realized without the cooperation and help of the farmers and officials of Hong Ha and Huong Nguyen communes, as well as the extensionists from the Rural Development Department of A Luoi District and Extension Centre of the province. It is a pleasure to thank all members of the research team for their cooperation and for sharing their experience.

The Mongolian team notes that their collaborative research is a pioneering effort in which voices and aspirations of women herders have been included in the study and in the co-management arrangements for better pasture use. They thank, *bayarlalaa*, Ronnie Vernooy, Liz Fajber, John Graham and Hein Mallee from IDRC for their valuable comments and timely contribution and support for the organization of the Second Social and Gender Analysis International Learning Stories Meeting in Ulaanbaatar in 2003. They also thank all the local women's groups and representatives of the local governments in Lun, Khotont and Deluin *sums* for their cooperation during the study.

1

Integrating Social and Gender Analysis into Natural Resource Management Research

Photo credit: Ronnie Vernooy

RONNIE VERNOOY AND LIZ FAJBER

THE SOCIAL NATURE OF NATURAL RESOURCE MANAGEMENT

In most regions of the world, the sustainable management of natural resources, including biodiversity, requires the involvement of multiple social actors or stakeholders. Stakeholder involvement refers to the active and meaningful participation of small farmers (both men and women), large farmers, entrepreneurs, local authorities, local groups, non-governmental organization (NGO) staff and policy makers in decision-making processes concerning the use, management and conservation of natural resources. This includes the analysis of problems and opportunities, the definition of research and development initiatives, and the monitoring and assessment of action and plans. It often also includes working together to reconcile conflicting or divergent points of view and interests. In particular, the active involvement of NGOs, local governments, grassroots groups and farmer associations is now a feature in many participatory, natural resource management initiatives.

In such an approach, it is imperative to address both the ecological and sociological aspects of natural resource (management) dynamics. This usually means looking at larger landscape units, such as, for example, a watershed or a micro-watershed, a community forest or rangeland. It requires dealing systematically with the changing and often complex interactions among components of a natural resource system or a production system, such as farming, fishing, forestry, herding, collecting edibles or combinations of these. It also requires considering the historical, socioeconomic and political forces that influence these interactions. These forces in turn are defined by such variables as class, gender, age and ethnicity.

Foremost, it implies learning from the women and men living in rugged mountainous areas, desert margins, stressed coastal basins and other marginal areas, who are struggling to make a living under often very difficult conditions. The key questions to answer are: How do these people construct and perceive what is happening in their community, watershed or region? How do they view what we call the management of natural resources? What is their interest in participatory action research processes and do they see them as a way to create more room to manoeuvre? Are local women farmers and fishers interested in joining professional researchers in a collaborative effort to analyse their situation and to design, test and assess new or adapted management practices?

These considerations lead to exploring such processes as the generation, distribution and use of knowledge. Of particular interest is the study of the social and gender relations and configurations that condition access, tenure, entitlements, claims and rights to natural resources, including the social dynamics of change, adaptation and resilience. It also raises the cultural and political nature of research methods and practices.

This book documents and reflects on an initiative that recognizes the steps that researchers are already taking to implement social and gender analysis (SAGA) research including questions of class, caste and ethnicity in natural resource management. It presents learning studies from six diverse research teams in the field. The teams are from India, Nepal, China, Viet Nam and Mongolia.

Natural Resource Use in Asia: Trends and Problems

Despite rapid industrialization and urbanization in Asia, most people remain directly dependent on a productive natural resource base for their livelihood. Unfortunately, pressures on this resource base are increasing. Urban-biased industrial development and non-locally managed international investments in export-oriented resource development are leading to degradation of those resources. Resettlement due to displacement, voluntary migration (mostly by men, such as in China) and historical conflicts exacerbate the pressures. Rural populations have increased rapidly because of improvements in basic health and living conditions. This leads to expansion of cultivated land, even into areas that are ecologically fragile or inappropriate for permanent cultivation. Within communities, marginalization processes are common. Systems of tenure and access to resources are complex, as traditional, culturally specific systems are modified by colonial and state regulations that may be changing rapidly with national economic policy reforms.

Problems related to the sustainable management of natural resources are most critical in the uplands and coastal areas, where natural resource degradation can often lead to irreversible loss of food sources and the breakdown of ecosystems with loss of habitat. In Asia there is widespread privatization of natural resources, such as forests and coastal areas, that were previously managed collectively. Privatization may lead to productivity increases in the short term, but it also frequently increases poverty because poor people who previously had access to these resources are now excluded. Conventional policies and research have often discounted the role of local people in the design and implementation of measures, projects and programmes, and are often blind to social differentiation.

Although circumstances differ in different countries, there is a striking convergence of interest in questions of local resource management. In some countries structural adjustment is leading to reductions in the technical and enforcement capability of the state. In others major policy transitions are affecting all aspects of government interventions in the economy. External pressures due to expanding trade and investment and large-scale development projects in parts of the region previously isolated from international markets are also having a dramatic effect on local resources. At the same time, local governments and grassroots organizations are becoming more assertive and articulate in their identification of resource questions—and the expression of *their* views and interests.

THE CHALLENGE OF INTEGRATING SOCIAL AND GENDER ANALYSIS INTO NATURAL RESOURCE MANAGEMENT RESEARCH

The complexity of societies in Asia and the problems of natural resource management are considerable. Notions of gender (the socially constructed roles and characteristics assigned to men and women in a specific culture), class, caste, ethnicity and age are integral to understanding the social relations and decision-making processes concerning access to, and use and management of natural resources. A sound understanding of social differences and social inequality are key to finding answers to the questions outlined in the previous sections. Simple answers are unlikely, as Kabeer (2003: 193) points out:

> Gender relations, like all social relations, are multi-stranded: they embody ideas, values and identities; they allocate labour between different tasks, activities and domains; they determine the distribution of resources; and they assign authority, agency and decision-making power. This means that gender inequalities are multi-dimensional and cannot be reduced

simply to the question of material or ideological constraint. It also suggests that these relationships are not always internally cohesive. They may contain contradictions and imbalances, particularly when there have been changes in the wider socio-economic environment.

Who participates in development (research) interventions, projects, programmes, and policies? How exactly? Who benefits from them? Who remains excluded or isolated? These are becoming crucial questions to be considered and integrated into intervention strategies if the aim is to support the more equitable—and sustainable—use of natural resources and the derived benefits.

Some policy makers, activists and researchers in the region recognize the need to reflect on and integrate social and gender equity, particularly as it relates to participation, inclusion and exclusion, decision making and power relations. Agarwal (2001: 1623) has forcefully drawn attention to processes of exclusion in the case of the formation and operation of community forest groups:

> Ostensibly set up to operate on principles of cooperation, such [community forestry] groups are meant to involve and benefit all sections of the community. Yet effectively they can exclude significant sections, such as women. These 'participatory exclusions' (that is exclusions within seemingly participatory institutions), constitute more than a time-lag effect. Rather, they stem from systemic factors and can, in turn, unfavourably affect both equity and institutional efficiency.

Studies such as Agarwal's improve our understanding of these key social and political processes informed by gender and other variables. However, the practical and context-specific implementation of more socially sensitive research and development interventions in relation to biodiversity and natural resource management remains a very difficult process for many. Most of the social and gender analysis in natural resource management is primarily at the conceptual level. There are few effective learning programmes that focus on systematic capacity building for gender and social analysis in applied research in this field. There are even fewer initiatives that systematically document and analyse this kind of capacity-building process.

The challenge of integrating SAGA into natural resources and biodiversity research are, therefore, many (Vernooy and Fajber 2004: 210):

1. Knowledge of and experience in social science research among natural resource management researchers and research managers is limited.
2. Social science components are not well integrated with natural science components in most research efforts.
3. Researchers and research organizations have different starting points, interests and expertise in terms of social and gender issues.
4. 'Gender blindness' or the refusal to acknowledge the importance of gender issues is common in research and research policy making.
5. Short-term training has limited impact.
6. Resources in the area of SAGA and natural resource management in Asia are not widely available.
7. Networking has potential benefits but operationally is not easy.

Integrating SAGA requires sound institutional analysis of how production and reproduction are organized at household and community levels and how these relate to (inform and are informed by) the market and the state.[1] Several recent studies show how this can be done effectively. By collecting a series of detailed case studies from around the world, Howard (2003) shows how gender relations inform biodiversity management and conservation, and why, in several cases, women predominate—particularly in the management of local plant biodiversity. In an example related to crops and biodiversity, Farnworth and Jiggins (2003: 5) note: 'One of the strong reasons why different men and women, and women of different backgrounds, have different [varietal] preferences is because they relate to the *food chain* in different ways, and often at different times and places.'

In summary, integrating SAGA into research is important in developing a better understanding and awareness of the social and power relations that govern access to, use of and control over natural resources. This involves understanding the differences and the inequities of social actors and is dependent on the local contexts.

> Shifting the focus from fixed identities to positions of power and powerlessness opens up new possibilities for addressing issues of equity. In practical development terms, this implies more of a role for participatory approaches to explore, analyze and work with the differences that

people identify with, rather than for identifying the 'needs' of pre-determined categories of people. This calls for an approach that is sensitive to local dimensions of difference and works with these differences through building on identifications rather than superimposed identities. (Cornwall 2000: 28–29)

It is also important for facilitating the recognition of the social and gendered nature of technologies, policies and interventions. Policies and technologies are value-laden; women and men and different social groups are involved and affected differently.

Gender-awareness in policy and planning requires a prior analysis of the social relations of production within relevant institutions of family, market, state, and community in order to understand how gender and other inequalities are created and reproduced through their separate and combined interactions. (Kabeer 1997: 280–81)

A last reason for integrating SAGA into research is to create space for social actors (women and men) to manoeuvre and to enhance the bargaining and negotiating power of marginalized and discriminated groups, leading to empowerment and transformation where they have more access to, control over and benefits from natural resources.

Home economists, health planners, agricultural planners, the environment lobby have all targeted women in their plans on the basis of narrowly defined perceptions of what women do. The problem is that women, particularly poor women, do simultaneously undertake many of these roles and responsibilities, often without pay; hence their longer hours of work. Development interventions, designed and implemented by individual sectors with very little coordination between them, generate conflicting demands on women's time and energy. Such interventions are either doomed to failure (thereby confirming planners' worst fears about women's irrational behaviour) or else result in the intensified exploitation of women's labour. (ibid.: 270)

THE 'LEARNING STUDIES' PROJECT

To address this situation and as a direct response to requests from our Asian research partners for more field-based training and exchange of practical experiences and methods in implementing SAGA in the field, two programmes of the International Development Research Centre (IDRC)—Community-based Natural Resource Management (CBNRM) in Asia and Sustainable Use of Biodiversity (SUB)—developed a novel 'umbrella' or multiple-component, capacity-building programme. The programme (Supporting Capacity Building for Social/Gender Analysis in Biodiversity and Natural Resource Management in Asia: An Umbrella Activity) was approved in 2002 and implemented step-wise (Vernooy and Fajber 2004: 209–10).

In Asia the diversity of cultures and languages reinforces the need for locally relevant methods and training approaches, a concern frequently expressed by IDRC partners. There is significant criticism that most methods and concepts are grounded in 'Western' thought and are not always applicable in the social and cultural contexts in which our partners are working. Therefore, the capacity-building programme strives to work with research partners to develop and adapt tools and methods to culturally relevant conditions, including language and learning examples.

The programme objectives are:

1. to support the integration and practical application of SAGA at the field level through training and support programmes;
2. to develop culturally appropriate (or adequate) approaches and tools for SAGA in natural resource management research;
3. to support interactive south–south networking and information exchange among researchers interested in integrating SAGA into natural resource management research;
4. to build capacity within institutions to mainstream gender in project activities and within the institutions themselves; and
5. to document best practices and progress made by researchers toward integrating SAGA into natural resource management research in Asia (process and outputs of objectives 1 to 4).

One of the activities implemented as part of the programme is called the 'Learning Studies' project. Initiated in 2002, this project recognizes the steps that researchers are already taking in implementing SAGA research in natural resource management and documents both the successes and failures that illustrate learning in this process. The project brings together six diverse research teams from five Asian countries representing both academic and non-academic sectors, a variety of research organizations, and researching a number of natural resource management questions, including biodiversity conservation, crop and livestock improvement, and sustainable grassland development:

1. Sikkim/West Bengal, India: The Eastern Himalayan Network (EHN). The EHN team includes Chanda Gurung, a gender and natural resource management specialist, and Nawraj Gurung, an extensionist by training, currently focusing on agricultural and horticultural issues.
2. Nagaland, India: The Nagaland Empowerment of People through Economic Development (NEPED) project. The NEPED SAGA team is formed by Chozhule Kiki, a social scientist with an interest in food and agriculture, and Vengota Nakro, a natural scientist specializing in agriculture and silviculture.
3. Nepal: Local Initiatives for Biodiversity Conservation and Development (LI-BIRD, an NGO). Deepa Singh, a horticulturist, Anil Subedi, a rural extensionist, and Pitamber Shrestha, a rural development specialist, make up the LI-BIRD team.
4. China: The Center for Chinese Agricultural Policy (CCAP) of the Chinese Academy of Sciences (CAS). Two social scientists represent the CCAP SAGA team: Yiching Song, with a background in rural development studies, and Linxiu Zhang, an agricultural economist.
5. Viet Nam: Hue University of Agriculture and Forestry (HUAF). The HUAF SAGA team is represented by Hoang Thi Sen, who has a background in forestry and agriculture, and Le Van An, an animal scientist. Both have a strong interest in rural development questions.
6. Mongolia: The Ministry of Nature and Environment (MNE) and the Gender Research Centre for Sustainable Development. The Mongolian SAGA team is represented by researchers from a number of organizations. Hijaba Ykhanbai and Enkhbat Bulgan work for the MNE. Tserendorj Odgerel is with the Gender

Research Centre for Sustainable Development, and Baatar Naranchimeg is studying at the Mongolian State University.

Projects are often criticized for weakness in SAGA and only very strong or nearly perfect projects that integrate SAGA are appreciated. In this initiative we recognize that learning is an iterative process. Through cross-regional exchanges, the project supports and encourages the steps along the path to learning. The selected cases illustrate real-world examples—in terms of challenges and opportunities, successes and disappointments—and highlight a number of methods used and adapted in the very diverse contexts of Asia.

The studies reflect on not only what has been done and is being done in organizations in terms of capacity development, but also how this has been done or is being done, and what enabling and constraining factors are affecting the process of integrating SAGA. In addition, they ask how best to support these capacity development efforts.

The case study approach is based on six guiding questions; some conceptual and methodological elements (such as an action-oriented approach); and an iterative process of workshops, fieldwork and the production of a number of outputs. The six cases also developed a common theory of action (Patton 1997) outlining how the research process could proceed (Figure 1.1). At the planning stage it helped the team think through the interlinked steps; at the end of the cycle it provided a means to reflect on the actual road followed.

The six cases are examples of pioneering efforts in the particular local context in which they operate; therefore, they are not to be confused with initiatives and results guided or headed by gender experts. Together, they reflect a diversity of strategies, approaches and methods. Some cases illustrate a 'women in development' (WID) approach and its defining features; others fit more within a 'gender and development' (GAD) approach where the focus goes beyond women and women's issues (Connelly et al. 2000: 140–48; Rathgeber 1994). Some cases combine elements of both or are moving from a WID agenda to a GAD approach (Table 1.1 lists the key features of both approaches). No two case studies are alike. We will come back to these approaches and elements in the concluding chapter.

Figure 1.1
Theory of Action

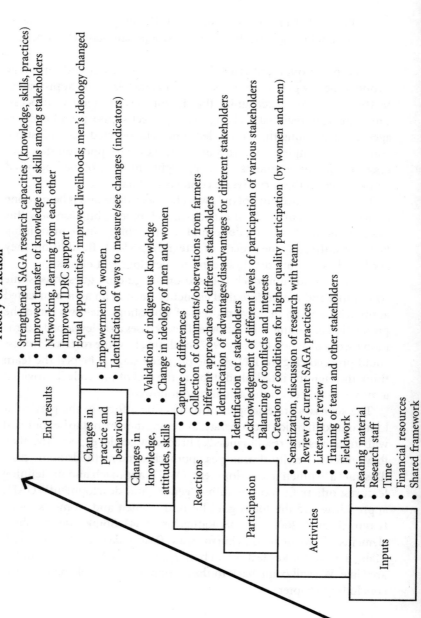

Table 1.1
Key Features of Women in Development (WID) and Gender and Development (GAD) Approaches to Research*

Characteristic	WID	GAD
Focus	Women and their exclusion from development initiatives	The socially constructed relations between men and women, and the subordination of women
Perceived core problem	Women's exclusion	Unequal power relationships
Goal	Women's inclusion and more effective development	Equitable and sustainable development, appropriate participation and decision making
Solution	Full integration of women in development process	Empowerment and social change
Main strategies	Women's projects; increasing women's productivity and income; increasing women's ability to look after the household	Reconceptualizing the development process taking gender and other inequalities into account; identifying and addressing practical needs of women and men; addressing women's strategic interests; addressing strategic interests of the poor and marginalized

Source: Adapted from Connelly et al. (2000: 141).
Note: * Some authors distinguish a third approach—women, environment and development (WED) (Leach et al. 1995). The WED approach has portrayed women as key users and managers of natural resources based on a special (nurturing) relation with nature. As none of the six case studies exemplifies a WED approach, we do not provide further details.

The six guiding questions, agreed on by the teams, are:

1. What does SAGA in natural resource management research mean for different stakeholders?

2. What are the key capacities required for different stakeholders to do SAGA?
3. How are these capacities developed and strengthened (for example, through networking, organizational support)?
4. What are the enabling factors? What are the constraining factors?
5. What have been the achievements of the work so far at different stakeholders' levels?
6. What more can be done to advance SAGA and social and gender equity?

THE 'LEARNING STUDIES' ROADMAP

Finding Common Ground: First Workshop

From 8 to 12 May 2002 participants from the six teams gathered for the first time in Beijing to share previous and ongoing research activities and experiences, to come to a shared understanding of the meaning of SAGA, and to plan concrete action for strengthening their SAGA efforts. Challenges and entry points for integrating SAGA into projects, programmes and organizations were identified. Each team prepared an action plan aimed at strengthening research on a particular SAGA question (in most cases as part of their ongoing research activities). While recognizing that social and gender problems depend on a specific context, common understanding was reached on the nature of the social and gender issues in the various projects. The main features of the six studies are summarized in Table 1.2.

The six studies each developed a particular set of methods and tools to respond to local needs and interests and to address the specific research questions. Together they form a rich and varied methodological basket that merits highlighting (Table 1.3).

In addition, agreements were made to continue sharing knowledge and experience. Steps towards networking were identified, including the use of the Virtual Resource Centre (VRC, a Web-based information and communication tool set up by the CBNRM programme), exchange visits and a second workshop planned for 2003. (A full report can be

Table 1.2
Key Features of the Six Studies

Case study	Focus	Methods	Major issues
Sikkim/West Bengal	Roles of men and women in ginger production and disease management	Analysis of engendered nature of ginger production	Cultural questions; migration, ethnicity and caste conflicts
Nagaland	Women in marketing	Engendered producer-to-consumer chain analysis	Economic change; population pressure; local politics
Nepal	Roles of women and men in seed production and marketing	Analysis of engendered nature of seed production; participatory rural appraisal; formation of seed producer groups	Formal research not very responsive to rural reality; new policies
China	Feminization of agriculture; formal research system disconnected from rural reality	Participatory plant breeding; social actor analysis; women's action groups	Rapid macro-economic changes; the state and location of ethnic minorities; new policies
Viet Nam	Women's roles in rural development	Women's needs and interests analysis; participatory rural appraisal; formation of interest groups	Cultural questions; macroeconomic change; the state and location of ethnic minorities; new policies
Mongolia	Roles of men and women in livestock management	Women's needs and interests analysis; participatory rural appraisal; formation of interest groups	Cultural questions; new policies and laws

Source: Case study research proposals.

Table 1.3
Methods, Tools and Skills Used in the Six Case Studies

Case study	Methods, tools and skills
Sikkim/West Bengal	Proposal writing skills (clear objectives and focused research questions); involvement of local research assistants (women and men); participatory rural appraisal (PRA) tools, historical analysis; communication skills; sharing experiences with other projects
Nagaland	Survey, interview, focus group, PRA tools; participant observation; negotiating skills; advocacy skills
China	Social and gender analysis action plan; collaborative action research (e.g., participatory plant breeding); strengthening of women groups; strengthening links between farmers (women), researchers and extensionists; policy analysis and policy experimentation; participatory monitoring and evaluation (PM&E); sharing experiences with other projects
Nepal	Literature review; PRA tools; cooperation with government
Viet Nam	Social and gender analysis action plan; collaborative research; formation of farmers' interest groups; PRA tools; PM&E; cooperation with extensionists; training skills
Mongolia	Literature review; PRA tools; qualitative methods; PM&E; organizing skills; facilitation skills; training skills; sharing experiences with other projects
All studies	Detailed documentation of process and results

Source: Case study proposals.

found at: http://www.cbnrmasia.org.) With the insights gained at the workshop, participants plunged back into their fieldwork.

Analysing and Comparing Field Experiences: Second Workshop

The second workshop was held in Ulaanbaatar, Mongolia, in October 2003, a few months later than originally scheduled because of the SARS outbreak, which slowed fieldwork and made travelling impossible for several months. Following the format of the Beijing workshop, a participatory approach was adopted with room for individual contributions in both plenary sessions and in small group work. Facilitation was provided by in-house resource people. The workshop was divided into four sessions: presentation of case studies and feedback (based on draft fieldwork reports aided by posters, photo albums and one video [from the Mongolian team]); identification of common issues; writing studies/stories; and planning.

KEY COMMON ISSUES

In reviewing individual case studies and synthesizing the most striking things and the missing or underemphasized elements, nine common issues emerged with respect to integrating SAGA into natural resource management. These elements, which together are providing partial answers to the six guiding questions, are: stakeholder analysis; gender roles: description and explanation; multi-stakeholder collaboration: initiation and fostering; diversified empowerment strategies; disempowerment; scaling up and scaling out; impact assessment (appropriate methods and tools); sustainability (ecological, socioeconomic); and systematic documentation.

In any given context, most, if not all, of these nine issues are interrelated. For example, the initiation of multi-stakeholder collaboration requires a sound stakeholder analysis. Developing an empowerment strategy for poor women requires an understanding of culturally defined gender roles concerning issues such as the division of labour; access to land, water, crops, and animals; access to services such as credit, training and extension; and the organization of women. Four of the nine issues (gender roles; initiation and fostering; diversified empowerment strategies; and impact assessment) were selected by a simple, individual vote as priorities for more in-depth discussion. This was decided based on what the studies had achieved so far, and whether and what kind of additional work would be required to improve quality and rigour.

STRUCTURE OF THIS BOOK

The six cases studies follow this introduction and form the heart of this book. The cases have a common pattern. Each starts with a description of the context of the study. This is followed by a brief explanation of the reasons for addressing social and gender questions, research objectives and research questions. The core of the case studies focuses on the findings of the fieldwork carried out to address specific research questions. The sequence of the chapters flows from South Asia to South-East Asia to North-East Asia, as follows:

- Chapter 2: The social and gendered nature of ginger production and commercialization: A case study of the Rai, Lepcha and Brahmin-Chhetri in Sikkim and Kalimpong, West Bengal, India (Chanda Gurung and Nawraj Gurung).
- Chapter 3: Strengthening market linkages for women vegetable vendors: Experiences from Kohima, Nagaland, India (Vengota Nakro and Chozhule Kikhi).
- Chapter 4: Enhancing farmers' marketing capacity and strengthening the local seed system: Action research for the conservation and use of agrobiodiversity in Bara district, Nepal (Deepa Singh, Anil Subedi and Pitamber Shrestha).
- Chapter 5: Empowering women farmers and strengthening the local seed system: Action research in Guangxi, China (Yiching Song and Linxiu Zhang with Ronnie Vernooy).
- Chapter 6: Creating opportunities for change: Strengthening the social capital of women and the poor in upland communities in Hue, Viet Nam (Hoang Thi Sen and Le Van An).
- Chapter 7: Herder women speak out: Towards more equitable co-management of grasslands and other natural resources in Mongolia (H. Ykhanbai, Ts. Odgerel, E. Bulgan and B. Naranchimeg).

The case studies are followed by a comparative analysis of a number of key common issues and challenges identified by the teams. This analysis, grounded in empirical research findings, is presented in two related chapters. Chapter 8 responds to guiding questions 1, 5 and 6 (listed earlier), with a focus on social and gender insights. The chapter compares the main underlying forces or factors that shape particular management practices and some of the emerging issues in terms of equity and environmental sustainability. It reflects on the similarities and differences in how social and gender relations inform natural resource management practices in the study sites. The chapter analyses the cases along a continuum from descriptive to transformative, with reference to the key features of the WID and GAD approaches mentioned in this chapter. Each of the six studies combines elements of this continuum.

Chapter 9 responds to guiding questions 2, 3 and 4, with a focus on the process of integrating SAGA in research and development. It summarizes the main achievements and remaining challenges of the research processes in terms of capacity building. Results are encouraging, but certainly not perfect or final. In most cases research is ongoing and new issues are emerging as well. The chapter concludes with the identification of some major challenges still to be addressed: organizational change and mainstreaming of social and gender research; enduring inequities and the empowerment of women; the improvement of the quality of participation; and dealing with macro forces. The two concluding chapters emphasize that natural resource management questions, whether addressed from a micro or macro perspective, are not social or gender neutral. At the same time, the case studies demonstrate that the processes that shape everyday management practices are never homogeneous by definition. This is a key empirical finding.

NOTE

1. Production and reproduction refer to the activities and strategies used by basic social units (families, households, kinship networks) to make a living and to guarantee the survival of the unit.

REFERENCES

Agarwal, B. (2001). Participatory Exclusions, Community Forestry, and Gender: An Analysis for South Asia and a Conceptual Framework. *World Development*, 29(10): 1623–48.

Connelly, P., T. Murray Li, M. MacDonald and J.L. Parpart (2000). Feminism and Development: Theoretical Perspectives. In J.L. Parpart, M.P. Connelly and V.E. Barriteau, eds, *Theoretical Perspectives on Gender and Development*, pp. 51–159. Ottawa: International Development Research Centre.

Cornwall, A. (2000). Making a Difference? Gender and Participatory Development. IDS discussion paper 378, Institute of Development Studies, Sussex.

Farnworth, C.R. and J. Jiggins (2003). Participatory Plant Breeding and Gender Analysis. PPB monograph no. 4, Participatory Research/Gender Analysis Programme, Cali.
Howard, P., ed. (2003). *Women and Plants: Gender Relations in Biodiversity Management and Conservation.* London: Zed Books.
Kabeer, N. (1997). *Reversed Realities: Gender Hierarchies in Development Thought.* London and New York: Verso.
——.(2003). *Gender Mainstreaming in Poverty Eradication and the Millenium Development Goals: A Handbook for Policy-makers and Other Stakeholders.* London: Commonwealth Secretariat; Ottawa: International Development Research Centre; Hull: Canadian International Development Agency.
Leach, M., S. Joekes and C. Green (1995). Editorial: Gender Relations and Environmental Change. *IDS Bulletin*, 26(1): 1–8.
Patton, M.Q. (1997). *Utilization-focused Evaluation: The New Century Text.* Thousand Oaks: Sage Publications.
Rathgeber, E.M. (1994). WID, WAD, GAD: Tendances de la Recherché et de la Pratique Dans le Champ du Développement (WID, WAD, GAD: Research Trends in Gender and Development). In H. Dagenais and D. Piché, eds, *Femmes, Féminisme et Développement* (Women, Feminism and Development), pp. 77–97. Montreal and Kingston: McGill-Queen's University Press.
Vernooy, R. and L. Fajber (2004). Making Gender and Social Analysis Work for Natural Resource Management Research: An Umbrella Program for Building Researcher Capacity. In *Gender Mainstreaming in Action: Successful Innovations from Asia and the Pacific*, pp. 208–23. Washington: InterAction's Commission on the Advancement of Women and Silang; Cavite: International Institute of Rural Reconstruction.

2

The Social and Gendered Nature of Ginger Production and Commercialization

A case study of the Rai, Lepcha and Brahmin-Chhetri in Sikkim and Kalimpong, West Bengal, India

Photo credit: Nawraj Gurung

CHANDA GURUNG AND NAWRAJ GURUNG

The Region

Sikkim is a small state of India, situated in the inner mountain ranges of the eastern Himalayas (area: 7,299 km^2; elevation: 300–8580 m above sea level). Kalimpong (area: 1,056.5 km^2), which was part of Sikkim before 1706, is one of the four subdivisions of Darjeeling district (area: 3,281.87 km^2; elevation: 300–3,000 m above sea level) in the state of West Bengal.[1] The two areas are adjacent to each other and fall within the eastern Himalayan agroclimatic zone. The climate of the region varies from cold temperate and alpine in the north-east to subtropical in the south. The mean annual rainfall ranges from 2,000 to 4,000 mm.

The region has three major ethnic groups—the Bhotias (descendants of Tibetan and Bhutanese immigrants who came to Sikkim in the 16th and 17th centuries), the Lepchas (the indigenous people of Sikkim) and the Nepalis (who migrated from Nepal in the 18th and 19th centuries). In addition, others came to Sikkim as businessmen, traders, service providers and labourers.

Agriculture is the mainstay of the region and 80 per cent of the people depend on it. Most farmers are smallholders, as per capita availability of land has been declining rapidly due to population pressure. The climate and seasons are conducive to growing a large number of high-value cash crops like cardamom, ginger, potatoes and horticultural crops. In general, rice, wheat and oranges are grown at lower elevations, while crops like maize, potatoes, ginger and cardamom are grown in the higher areas. Because of the favourable climate, many people are also commercial flower producers.

Numerous ethnic groups, with their own traditional cultures and languages, inhabit the region. However, as a result of long interaction, most of them have integrated to a considerable degree. Many practices and beliefs and even terms from different languages are now used commonly. The most evident sign of this integration is the use of the Nepali language by almost everyone, especially outside the home. Table 2.1 contains demographic information about the study areas.

40 Chanda Gurung and Nawraj Gurung

Figure 2.1
Location of the Study Sites in the Sikkim Region

Source: Perry-Castañeda Library Map Collection.

Table 2.1
Demographics of the Two Study Regions Compared with National Data, 2001

	Kalimpong	Sikkim	National
Total population	225,143	540,493	1,025,251,059
Males	115,521	288,217	530,422,415
Females	109,622	252,276	494,828,644
Urban population	42,980	60,005	284,995,688
Rural population	182,163	480,488	740,255,371
Population density	413/km^2	76/km^2	324/km^2

Sources: Government of Sikkim (1993); Cultural and Information Office, Subdivision Office, Government of West Bengal, Kalimpong, District Darjeeling (office files).

ETHNIC GROUPS

Lepcha

Lepchas are the indigenous peoples of Sikkim and Kalimpong and are little known outside the eastern Himalayas. Several hypotheses have been suggested regarding their place of origin: some claim Tibet (Pinn 1986), while others have cited possible links to the Kirats of east Nepal (Fonning 1987). There is no accepted theory or accurate knowledge of this. The indigenous account of the Lepchas does not contain any tradition or history of migration; they believe that they originated from the divine peaks of Mount Kanchenjunga and the valleys around them. They refer to themselves as *mutanchi rongkup*, the beloved sons of the Mother Creator.

The Lepchas have been influenced largely by two other major ethnic groups found in the region—the Tibetans, who ruled Sikkim for about 300 years, and the Nepalis, who migrated to Sikkim and Darjeeling district in the late 18th and 19th centuries. The Tibetans were powerful then, as they were the rulers, while the Nepalis, who migrated to this area in large numbers, became socially dominant in many ways. In addition, there was the influence of the Christian missionaries, who arrived from the nearby British-ruled Darjeeling area.

Rai

The Rais are an ethnically distinct but heterogeneous group of people belonging to the Tibeto-Burman subfamily. Physically, linguistically and, to some extent, culturally, the Rais are related to the Tibeto-Burman peoples, who spread eastward throughout the sub-Himalayan regions and the hills of Assam. At the time of their arrival in east Nepal (Limbuan territory), the Rais still retained their essential tribal characteristics (Chemjong 1966). They had successfully avoided the Sanskritization of Hindu influences from the south and Buddhist beliefs from the north. However, after the 'unification' of Nepal in the mid-18th century—more specifically from 1787, when the Gorkhas of Nepal conquered the Limbuan territories of the Rais and Limbus (Basnet 1974)—these people were integrated into Nepal.

Facing numerous challenges to their traditional way of life, they underwent tremendous transformation in their religious, sociocultural and subsistence systems. The Rais migrated to Sikkim and the Darjeeling area in the early 19th century to work in the tea gardens and to settle in uninhabited areas, clearing the forest and seeking income opportunities. By this time the Rais had become 'Nepalis' and more Hindu-like, losing much of their traditional tribal characteristics.

Brahmin-Chhetri

These two groups belong to the Khas community of the Indo-Aryan group. The Chhetris are historically the rulers of the Gorkha kingdom and the Brahmins are one of the castes (the other castes are Thakuris, Kamis, Damais and Sarkis) who are subjects of the Chhetri kings (Subba 1992). Both groups are Hindu and have made up the ruling upper class in Nepal. In the early 19th century, after the Indo-Nepal and Sikkim-Nepal wars, they began migrating to Sikkim and the Darjeeling area. They appeared in Darjeeling especially after the Treaty of Titalia in 1817 between Nepal and British India when the British began to enrol them in their army and settle them in uninhabited areas to work on the land as labourers and peasants. Their arrival in Sikkim began after 1861, following Sikkim's defeat by the British the year before. (Although the British defeated Sikkim, they did not annex the area; Sikkim became a protectorate [Rao 1978].) The British encouraged their immigration and settlement in the unoccupied regions of Sikkim as they realized that they 'made as a good a peasant in peace-time as... a soldier in war' (Basnet 1974: 42). The Brahmins soon followed suit and immigrated to this region. They were brought along as priests, especially by the Chhetris and Thakuris.

INTEGRATING SOCIAL AND GENDER ANALYSIS INTO GINGER PRODUCTION AND MARKETING

Social and gender relations are a result of a complex array of factors that depend on social, cultural and historical circumstances. Social and gender systems are linked to religion and ethnic identity, but at the same time are strongly influenced by external forces on the communities. Thus, social and, more particularly, gender relations are frequently reinterpreted and renegotiated as traditional societies are faced with new influences and symbols of change (Gurung 1999). To understand and analyse the social and gender relations within and among the three ethnic groups in this study, it has been crucial to delve into the historical background of the region and its peoples. A dynamic and complex picture emerges from our study.

In any community, gender representations are not uniform; rather, there are discontinuities and contradictions. The lives and activities of individual women express how they selectively embrace, tolerate, oppose or ignore gender ideologies. In addition, individual women and men, depending on the situation, behave differently in different activities and spheres of life. Thus, the individual activities and practices of women and men demonstrate how the wider forces of change are manifested at the local level through individual lives. However, the importance of individual choice and action is frequently overlooked when highly abstracted views of society assume a homogeneous and unchanging social order that forms an idealized situation often informed primarily by male perspectives (Watkins 1996).

Ginger is the main, if not only, cash crop for many farmers in our study area. Although cardamom and tea are also important cash crops, they require more specific conditions and relatively large areas to cultivate them commercially. Both crops, especially tea, also require processing before they reach consumers. In contrast, ginger can be grown economically on small plots in a wide range of environments. Thus, smallholders and marginal farmers can grow ginger and sell or consume the crop without any processing. Ginger is an annual and rhizomes can be harvested twice a year—the main rhizome halfway through the growing cycle and new rhizomes when the crop is fully grown. Because farmers rely so much on this crop, control and decision making in the production cycle are of great importance.

Significant changes are occurring due to commercialization.[2] Traditionally, the Rai community was the major grower of ginger, although the Lepchas also cultivated ginger for use in religious ceremonies. Both communities practised traditional production methods. However, as the crop became more important commercially, other communities, especially the Nepali Hindu Brahmin-Chhetris, took up its production. These communities had their own beliefs, practices, technologies and methods of cultivation, and gradually their production practices had an affect on the Lepchas and the Rais. At the same time, the commercial value of ginger increased significantly. More non-traditional areas were brought under ginger cultivation, and new techniques were introduced and practised. Gender roles in cultivation, decision making and control over the crop also changed.

These transformations led to some important changes. Over the last 15 to 20 years diseases have affected the crop resulting in a decline in the yield ratio from 1 'seed' rhizome to 8–10 harvested rhizomes to only 1 to 2–3. Many farmers have given up ginger cultivation and others are still struggling to survive because of the absence of alternative income-producing crops. Incomes and living standards have declined.

As the commercial value of ginger increased, the Brahmin-Chhetri communities began taking over its cultivation with their more advanced methods and new technologies, and the Rais and Lepchas lost their traditional control over the crop. Social relations between the communities began to change. The Brahmin-Chhetri who migrated from Nepal have been considered more entrepreneurial as they have always been more involved in agriculture than the other two groups. With their social connections and wider exposure to new information and new agricultural technologies, they soon surpassed the Rais and Lepchas in terms of landholdings and they became the dominant group in the social structure. The Rais and Lepchas gradually began to adopt their practices and beliefs, although with some adaptations.

The Brahmin-Chhetri domination of ginger cultivation meant that women's roles in production and control over ginger declined drastically. For example, the Hindu belief in the 'impurity' of women bars them from entering agricultural fields or even touching the crops. This was compounded by the strongly defined gender roles with regard to 'outside' and 'inside' work among Hindus. They consider any form of monetary function as 'outside work' and, therefore, in the men's sphere. Thus, ginger—a major cash crop—was brought completely under the control of men.

Several research stations and projects are hoping to find solutions to ginger disease problems. Surveys concerning farmers' knowledge and perceptions of diseases and cultural practices have been conducted. Based on the results of these surveys and research, extension messages have been developed. However, this research and extension work lacks any social and gender analysis and is strongly male biased—a serious shortcoming.

STUDY OBJECTIVES AND RESEARCH QUESTIONS

This study focuses on the social and gendered nature of ginger production, commercialization and disease management. The study is important for a number of people. First, for researchers—to allow them to understand social and gender dynamics, and to identify the constraints and opportunities of the various social groups regarding their livelihood and the role of ginger. Second, for research station staff and extensionists—to allow them to understand social and gender dynamics, and how these affect practices (especially in relation to disease management) which could help them come up with methods more acceptable to the farmers. Third, for farmers—ultimately, to help them examine their own methods and practices and come up with solutions for disease problems themselves.

The study also aims to explore strategies for involving women in the management and control of ginger production. Such strategies for addressing women's practical interests would, we hoped, lead to positive changes in their livelihoods. In particular, we set out to identify the enabling and constraining factors affecting the more disadvantaged groups in the region (women, poor, lower castes and classes) concerning control over ginger cultivation as a cash crop.

The following research questions guided the work:

1. What are the roles and responsibilities of men and women in the three ethnic groups concerning the cultivation and commercialization of ginger?
2. How does decision making take place among men and women in the three different ethnic groups concerning access and control over the key resources (land, labour, knowledge, capital)?
3. How do more general transformations in social and gender relations and in society affect the decision-making process for an important activity such as ginger production?

To the Field

We began our research with discussions with the coordinator of the Eastern Himalayan Network (EHN) who knows the region well. This led to a discussion of the importance of ginger to the Rai, Lepcha and Brahmin-Chhetri people.

The Centrality of Ginger

Rais. For the Rais, ginger is a very important crop that is required in all religious ceremonies. Even before the crop became commercial, they cultivated it in small areas near the homestead. The farmers follow and observe various taboos and practices, the most important of which is that non-family members and all animals are not allowed to enter the ginger fields. In fact, in some families only one person, usually the father, is the only one to enter the area. It is believed that spirits reside in the ginger fields and outsiders entering the field would be cursed by it, resulting in a disease that would make his or her limbs distorted and swollen (*dewa*). When ginger is harvested, the families perform a religious ritual called *naya ko puja*, in which the new crop is offered to the gods and spirits. It is only after this ritual has been performed that the ginger crop can be brought inside the house and used.

Lepchas. The Lepchas respect ginger and give priority to the crop as it is required in all their religious rituals and also for curing diseases.[3] All

Lepcha households cultivate ginger because they are not allowed to take ginger from others for rituals. Thus, among Lepchas and Bhotias, ginger remains a relatively important crop. Like the Rais, the Lepchas observe a religious ritual before using, eating or even bringing the new ginger crop into the house. In this ritual the spirit from the river Teesta is called on and offered paddy, a red cock and some local wine along with ginger. Once the offerings are made and the chanting completed, the spirit is guided back to its original place. It is believed that if this ritual is not observed, bad things will happen to the family and household; for instance, people will fall sick, cattle will die, or crop production will be poor.

Brahmin-Chhetris. These people, who settled in the region in the 19th century, took up the cultivation of ginger as the crop became commercially important. Because the crop is not religiously or culturally important to them, they adopted new technologies and methods to increase yield based mainly on commercial interests. This has resulted in cultivation and other practices distinct from those of the other two ethnic communities.

Site Selection

We looked for sites where ginger was an important crop and where one of the three ethnic groups—Rais, Lepchas or Brahmin-Chhetris—dominated. We wanted sites that were not too scattered and where farmers would be willing to participate. Using these criteria, the team drew up a list of possible sites in consultation with local extension officers of the agriculture department, research stations and milk cooperative unions. Seven villages were chosen and visited. Based on information collected about each village, two villages in Kalimpong and one in Sikkim were finally selected.

Kharka-Sangsay. This village belongs to the Sangsay panchayat in block I of Kalimpong subdivision. Rais make up 75 per cent of the households (Table 2.2). Ten to 15 farmers are landless here. The nearest town is Kalimpong, which is 8 to10 km away. According to one of the farmers, Kharka-Sangsay belonged to the Bhotia people who used it for grazing cattle. Some Rai people in this farmer's family would also bring their

Table 2.2
Size and Ethnic Composition of Households in the Selected Villages

Study site	Village	No. of households	Dominant ethnic group (households)	Other households
Kalimpong	Kharka-Sangsay	120	Rai (90)	Chettri, Pradhan, Tamang and Mangar
	Nasey	83	Lepcha (53)	Tamang, Pradhan and Chhetri
Kalimpong	Peshore	117	Lepcha (45)	Scheduled Castes* (14 or 15) Limbu or Subba (5 or 6) Rai (3 or 4) Chettri (5)
	Tashiding	40	Lepcha (23)	Brahmin and Pradhan
Sikkim	Central Pandam (Karnithang, Boorong and Sajong)	500	Brahmin-Chhetri (400)	Bhotia (30) Rai (30) Biswakarma (lower Hindu castes) (7) Tamang (10)

Note: * Under the Indian constitution certain castes in the Hindu communities that are socially backward have been listed in the Schedule of the Constitution and thus they are known as Scheduled Castes. They are considered lower than the other Hindu castes.

cattle to graze. About four generations ago, these Rai people bought the land and settled on it. Gradually, more people arrived and today it is a relatively large Rai village.

Much of the land was swept away in a major landslide in 1968, and only 30 per cent is now available for cultivation. In the non-cultivable area people grow trees for fodder and broom grass. As a result of this, cattle farming has improved.

The major crops are ginger, flowers, broom, large cardamom, vegetables and maize. Villagers tried to grow rice, but did not succeed because the soil does not retain water. All the households depend on farming, but their operations are very small. All of the farmers and households cultivate ginger, which is their major, or only, cash crop.

Nasey, Peshore and Tashiding. These three villages in the Kalimpong area are adjacent to one another (Table 2.2). The nearest town is Kalimpong, approximately 8 to10 km away. As in the other villages of the area, most of the people depend on farming and are all smallholders. The main crops are rice and maize, but they also cultivate round chilli, ginger, cucumber, tomatoes and other vegetables. According to the residents, about 15 years ago 40 to 45 per cent of the total cultivated area was devoted to ginger, but ginger now is grown on less than 5 per cent. Most of the farmers now cultivate this crop for home consumption only. The main reason for this drastic reduction in ginger cultivation is disease.

Central Pandam. Central Pandam is a *gram* panchayat (the smallest governing unit) in the East district of Sikkim. It has six wards, but only three were selected for study: Karnithang, Boorong and Sajong. Of the approximately 500 households in these wards, 400 are Brahmin-Chhetri (Table 2.2). The Brahmin-Chhetris, Bhotias, Rais and Biswakarmas are Sikkimese people, but the Tamangs have mostly come from Nepal. They cultivate leasehold lands, mostly on an annual payment basis. All the households are dependant on farming. About 300 Brahmin-Chhetri households and almost all the households of other castes grow ginger for a total of approximately 400. The other major crops are maize, mustard, finger millet, buckwheat, broom and vegetables. The nearest town is Rangpo, at a distance of about 5 km.

The Households

Several criteria were used for selecting the households and farmers: wealth, head of household (female, male, jointly headed), young and innovative, religion, length of time in the area, size of farm. In consultation with key informants, a relatively small cluster of 15 farmers representing different categories of households from each site was selected. To get a deeper and personal insight and perceptions of the ethnic groups, two field staff—one male and one female—from each of the sites and ethnic groups were selected to help collect information and data from their own village and community. The field staff were trained in communication methods and in conducting semi-structured interviews.

Tools and Methods

Different tools were used at the various stages. Various participatory rural appraisal (PRA) tools were used for data collection, such as a seasonal calendar, activity profile, semi-structured interviews, group discussions, key informants and on-site observation. We collected several oral testimonies. A review of secondary sources (books, journals, papers, articles) provided historical and background information. Attempts were made to collect information about when the crop was not commercially important in the area, when the crop became commercially important, when the crop began suffering from diseases, the period before the Brahmin came to the region, after the Brahmin settlement and start of ginger cultivation, and the present status. Data analysis focused broadly on social and gender roles, and relations with reference to ginger cultivation, diseases and management. We paid special attention to variables such as key stakeholders, economic relations, markets and institutional linkages.

UNDERSTANDING GINGER PRODUCTION AND COMMERCIALIZATION

Local History of Ginger Cultivation at the Three Sites

Kharka-Sangsay. The Rais have been living in this area for four generations and other groups have arrived gradually. In the early days farmers used traditional cultivation methods, applying cow dung and compost to the whole field, mixing it with the soil, then planting ginger without making rows or beds. The area was fenced to exclude animals and poultry. The farmers of Kharka-Sangsay grew the *nangrey* variety of ginger (literally, 'with nails'; a claw-like variety).

People started cultivating ginger on a larger scale in about 1965–66 when they learned of its commercial value from the Rais of Gorubothan (a town located 65 to 70 km east of Kalimpong). The farmers began to get 'seed' ginger from Gorubothan farmers—the variety is still known as *gorubothaney* (meaning 'from Gorubothan'), and this variety has now completely replaced *nangrey* ginger. Farmers prefer it as it is bigger and fetches a good price in the market. About six years ago they began to notice diseases in the ginger.

The farmers believe that the situation in Kharka-Sangsay is much better than in other ginger-cultivating areas, mainly because of the climatic conditions—the village is quite shady and cooler than other production areas. However, these growing conditions do not seem to be favourable for production of rhizomes that can be used as seed. The local seed is not good quality, and farmers must buy seed every year from neighbouring villages.

The Rais, who have been 'Sanskritized' for some time, observe several Hindu practices. For example, menstruating women are not allowed to enter the ginger fields for 10 to 12 days each month in the belief that at this time their blood contains harmful germs that might be carried by air to the ginger plants, thus causing diseases. The Rais also used to observe the Hindu tradition of not working in the fields during *saran* which lasts from two days to a week, depending on the moon and comes approximately once in 27 days according to the lunar calendar. In addition, when there is a death in the family, members do not work in or enter the fields for 13 days, equivalent to the mourning period. When there is a birth in the family, the Rais observe a seven-day ban on working with ginger as the family is considered to be 'impure' during that period.

Peshore, Nasey and Tashiding. Before the farmers began large-scale cultivation of ginger, the only fertilizer they used was ash, which was spread over the field. The men would plough the land while the women planted the crop. They collected seed rhizomes from the current crop and stored them in pits in the field for sowing the following year. Old Lepcha farmers point out an interesting fact: previously, other communities did not buy seed from the Lepchas, Rais and Limbus because they were afraid of the spirits living in the ginger fields of these communities. They believed that if they bought ginger from them, they might catch the *dewa* disease.

In this area commercialization of ginger began among the Lepchas only in about 1979–80. The Lepcha farmers saw how other farmers, especially the Tamangs of Peshore village, earned a lot of money by selling ginger and began to cultivate it commercially. During the early 1980s 50 per cent of the Lepcha farmers of the area were growing ginger for sale, although most of the older farmers did not. Disease set in very soon and commercial cultivation declined rapidly. Currently, only 10 per cent of the people in this area cultivate ginger on a commercial basis.

Disease became severe in the mid-1980s. According to some farmers, one of the main reasons for its rapid spread is that when ginger cultivation started commercially there was not enough seed and they had to bring some in from other areas. They believe that these rhizomes carried the disease. Farmers also purchased from Muslim traders who used to visit the area. The farmers also have other theories about the causes of disease problems. One is the use of fertilizers; as the farmers began to cultivate ginger on a large scale, they began to apply fertilizers to increase yield. Larger-scale production meant disease problems. Second is the belief that water flowing to the village from the Durpin area, an army cantonment, is contaminated with various chemicals like soap, oil and petrol. Third is deforestation. Fourth is the short time between cultivation periods; farmers feel that soil should be given a rest every five to seven years, a practice that earlier farmers followed.

Central Pandam. According to the oral testimonies of farmers 75 to 83 years old, the community settled in this area approximately a century ago. A family belonging to the Pradhan community brought them here to clear the land and settle. The area was originally a forest. The old men remember that during those early days people used to grow only 2 or 3 kg of ginger for home consumption. No one thought of selling it. The old farmers remember their mothers doing the work. They remember the *dhami* (shaman) asking regularly for ginger, which he needed to perform his rituals.

Ginger began to be cultivated on a large scale in the village in the 1960s. The first person to do so was from the Dadiram family; he had obtained 80 kg from somewhere (the farmers do not know or remember from where). Mr Dadiram used to travel to other villages frequently where he had seen this crop being cultivated in large amounts and sold as well. At first, he would grow ginger only once in three years and sell it in Siliguri. Soon other farmers also started to cultivate ginger in larger quantities and sell it at the local market and in Rangpo. Commercialization became widespread about 20 years ago. Before that there was no cash crop in the area. Disease was seen for the first time about 15 or 16 years ago. According to the farmers, the reasons for this were loss of soil fertility due to overuse; the purchase of seed rhizomes from various places; and because people have abandoned traditional practices, taboos and rituals.

This orthodox Hindu community observed a series of taboos and practices based on rituals and beliefs. Apart from the belief in the impurity of women, they used to observe *saran* and not work in the fields (especially sowing and harvesting). They also did not start planting or harvesting on the days that ended with '*si*' (for example, *ekadasi*) and '*mi*' (for example, *saptami, austami, nauwami, dasami*, etc.). Similarly, they used to avoid working in the fields on Tuesdays for religious reasons. The community did not work in the fields when there was a death in the family or at the time of birth. Nor did they work on the *tithee*s (the anniversary of the death of family members).

Inter-ethnic Social Relations: Some Key Points

Historically, the region and its people have been exposed to various external influences from historical movements of people due to conquests and wars, trade, migration and immigration. They have been confronted with a myriad of influences: the lamaist culture of Tibetan Buddhism, the caste system of Hinduism, the more 'egalitarianism' of Christianity, British colonialism and lowland traders. The process of interaction with, and influence of, these external hegemonies was based on relations of inequity. Stratification on the basis of class, caste, wealth, religion and gender became the norm.

A study of the social relations among the three communities shows how they have influenced each other. The Rais, for whom ginger is a very important crop religiously, and who were the first to migrate to and settle the region, had close contacts with and an influence on the Lepchas. Both ethnic groups belong to the Tibeto-Burman community and share many traditional tribal beliefs and practices. Intermarriage has reinforced exchanges. This is a major reason for the similarity in the religious rituals surrounding ginger. According to the oral testimonies of the old

Lepcha farmers, they learned to do the *naya ko puja* ritual from their daughters who were married into the Limbu ethnic group. The Limbus and the Rais belong to the Kirat group and share traditional beliefs, rituals and practices. When the Brahmin-Chhetris settled in this region, their influence began to be felt by the Rais and Lepchas. As the old Lepcha farmers said, they learnt to plant ginger in beds from them. But the Brahmin-Chhetris also learned ginger cultivation methods from the Rais and the Lepchas; the practice of storing ginger underground is an example. They also began to observe the *naya ko puja*.

Commercialization of the crop resulted in closer social relations among the various communities. The farmers began to exchange ideas and information about cultivation and to trade, buy and sell seed. Market forces deepened social relations, not only among the three main ethnic groups, but also with other groups. Traders from the lowlands represented market forces; they bought ginger and set the price. Many of the farmers also began to buy ginger seed from these Muslim tradesmen, creating a new social division: farmers (no matter to which ethnic community they belonged) versus those in control of commercialization.

However, differences remained, at least in people's perceptions. The Lepchas are considered more reticent and less sociable, while the Brahmin-Chhetris are more enterprising, better agriculturists and have more knowledge. The Rais fall between these two groups. Today, social relations among all the ethnic groups continue to be dynamic with a lot of give and take and exchange of information on cultivating methods, disease management, prices, places to sell and sources of seed. However, in most cases, the *parma* system[4] remains within the same ethnic community.

Current Cultivation Techniques

In the selection of land for ginger cultivation, fields where water stagnates are avoided. Farmers also believe that planting the same land every year does not produce a good crop. Chosen fields are cleared of all vegetation by burning and the ash is used as fertilizer. In Nasey, Peshore, Porbung, Tashiding and Central Pandam the land is then ploughed, dug and hoed. A second plowing occurs 15 to 18 days later. Cow dung (8–10 cm) is applied to the fields and they are plowed again to mix the manure into the soil. In Kharka-Sangsay the terrain is too steep to plow; instead, people turn the soil by hand and hoe. Beds 20 to 25 cm high are prepared with drains between them. Just before planting, the stored ginger seed rhizomes are sorted again. Good healthy rhizomes, that is, large, shiny, free from spots or marks, and free from bud or eye injury, are selected. The rest are sold in the market.

The selected ginger rhizomes are planted in two or three zigzag rows in the beds and covered with up to 8 cm of soil. The spacing between the rhizomes should be about 15 cm. Some farmers believe that to increase production, one can plant three or four rows in a bed. During planting the seed rhizomes are broken into pieces to ensure that each has two to four sprouts. Immediately after planting, the beds are mulched with dry leaves up to 8 to 10 cm thick. Some farmers use mulch made by putting grasses in the cowshed for a few days so that it mixes with cow urine and dung. Mulching protects the seedlings from rain, prevents weed growth, keeps the soil soft and moist, and accelerates growth. Most of the farmers, especially in Central Pandam, practise intercropping with maize as maize plants provide shade to the ginger. Maize is planted in the drains in between the beds. The farmers prefer the improved 'NLD' variety as it is shorter and has less chance of being blown over by wind.

After about a month, the beds are weeded. Ginger shoots are still very small and the chances of damage to them are high. Weeds are uprooted and disposed of outside the bed. When the ginger plants have three or four leaves, the *mau* or mother rhizome is removed, although if it is too small, it is left in place. Care must be taken not to damage the roots and cause infection. The roots are immediately covered again with soil. The *mau* is sold wherever the farmers can get a good price for it; some sell it to traders at the nearby markets. In Central Pandam some farmers sell it to the cooperative society. This income is very important, as it solves financial problems during the rainy season. Fifteen days after *mau* extraction, manure is applied.

In Central Pandam, where the farmers keep the seeds, they select healthy looking plants, that is, plants that are not dried, which they harvest 15 to 20 days before planting and stored in a dry, shady place. The rest of the crop is harvested for sale. The farmers of Nasey, Peshore, Porbung and Tashiding also keep seed rhizomes, but they harvest the entire crop, then select healthy rhizomes and store them either in a pit covered with straw or in a cool, dry place in the house or godown.

In Kharka-Sangsay the farmers cannot harvest seed rhizomes, as mentioned earlier. The seed they buy from the neighbouring village is stored either in the house in a cool, dry place covered with gunny bags or outside in a dry, shady place. The ginger for sale is harvested depending on the market price. Farmers take the harvest to the market or sell it to traders. Those who keep seeds usually change them every two to three years to reduce the prevalence of disease.

Farmers have been experimenting with various disease management practices in different villages. Some Lepcha farmers clean all the mud from the bottom of the diseased plant and expose the roots to the sun. Rotten plant roots are scratched and ash is applied. Some believe that just exposing the roots of infected plants to the air and sun can stop further spread of disease. The farmers say that these roots or rhizomes germinate and grow well. Some farmers cultivate ginger only in sloping fields using traditional methods, that is, without much land preparation. They just scratch the soil to make a hole and plant the rhizome seed in it. Many farmers contend that red soil helps control disease. One Rai farmer planted *bikphul* (*Glorisa*) in a ginger field where disease had been rampant. According to him, his ginger crop was very good that year without any disease. However, he has not been able to verify or repeat the experiment because he has only a limited area for ginger cultivation. One Lepcha farmer planted some ginger on rock covered with soil. The crop here did not have disease.

Gender Roles and Ethnicity in Cultivation Practices

1. *Site selection* is done by the head of the family, whether male or female, of all households at the two sites at Kalimpong. Also, if a male head of household is absent for whatever reason, the wife automatically takes on the role of head of the family and is responsible for this task. In Central Pandam, however, it is mainly the male head of the family who does the site selection. Only when the male head was absent did women take on this task.
2. *Land preparation* is done equally by men and women. In Central Pandam and in Nasey, Peshore, Porbung and Tashiding men

do the plowing while women do the hoeing and digging; whereas in Kharka-Sangsay, where there is no plowing, both men and women are equally involved in digging and hoeing the fields. However, in cases where the families follow the *parma* system, it is the male who does this work as it is felt that men can work more than women. Similarly, if the households need hired labour for this work, they hire men.

3. *Sowing and planting* are done by both men and women. However, usually only family members are involved. The reason given for this is that hired labour will not plant with as much care as household members.
4. *Manure application* is done by both men and women equally.
5. *Mulching* is done mostly by men, although women help when required.
6. *Weeding* is done by both men and women, although when hired labour is required for this task, women are employed. The wages for female labourers (INR 25) are lower than those for men (INR 40).[5]
7. *Extraction of* mau *rhizomes* is done by both men and women. However, usually only family members do this work as much care is needed to keep from damaging the plant. Farmers prefer not to employ hired labour for this.
8. *Soil covering* is mainly done by men, as it is heavy and strenuous work requiring physical strength.
9. *Harvesting* is done by both men and women equally and usually only household members are involved.
10. *Seed storage* is done both by both men and women.
11. *Marketing* is done entirely by men, except in female-headed households.
12. *Purchasing seed* is generally done by men. However, where the head of the household is a woman or the male head of household is absent, then the female head of the family takes on this responsibility.

It is evident that in all the communities the roles of men and women at the various stages of cultivation are almost similar. The differences depend more on household and family circumstances than on ethnicity.

It should be noted, however, that Brahmin-Chhetri women in male-headed households have less say in matters of site selection and purchase of seeds than women in the Lepcha and Rai communities. As one of the old Brahmin farmers from Central Pandam put it, 'I do not think that women are experienced in selection of fields, so I do not trust them to do this.'

The oral testimonies of the older men gave us an historical perspective. We learned that among the Rais, men used to play a more dominant role. In most families the male head of household was completely in charge and he was the only one who entered the ginger field. Among the Lepchas, both men and women seemed to have had an equal role. As one of the old farmers explained, 'Men plowed the field and the women planted the seeds and covered them with soil.' The women did the seed selection. However, men would perform the religious rituals with women merely participating passively. The tradition among the Brahmin-Chhetris was for women to cultivate ginger when the crop was grown in small quantities for household consumption. However, with commercialization and the production of large quantities, the work was taken over by men with women merely acting as 'helpers'.

Other differences between traditional and current practices also exist. For example, not all Rais and Lepchas observe the traditional religious ritual before eating or using the new ginger crop. Those who do still observe it shorten the ritual by just offering ginger to the spirit. The Rais have discontinued the practice of not working in the fields when there is a birth in the family. The main reason given for this was that nowadays families are scattered and it is not always possible to get news on time. Farmers also said that this practice is no longer practical in the face of high labour demand and high salaries. The Rais have stopped observing *saran*; they think it is no longer either practical or feasible. The younger generation does not even know when *saran* occurs. Many Brahmin-Chhetris have also abandoned the practice. Similarly, the practice of not working on Tuesdays or '*si*' and '*mi*' days has become less and less common, and the ban on working in the fields on the anniversary of the death of a family member is practised by only a few farmers. However, in contrast, some of the Brahmin-Chhetri farmers have started to observe the *naya ko puja*, the ritual conducted before eating the new ginger crop.

GENDER RELATIONS AND ETHNICITY

We looked at four aspects of gender relations: the roles of men and women in the agricultural work; access to land, labour and capital, and limitations on access; decision making and control; and image and self-esteem. These last two are based on elements of cultural ideology, symbols and statements that explicitly devalue women and socio-cultural arrangements that exclude women from participating in various religious activities or from holding power in society. Thus, the image and self-image of women and their confidence are influenced by gender ideologies, state-ascribed roles, cultural taboos and expectations, education and exposure to the outside world, ability to earn money, roles in decision making, and their own inner sense of autonomy, identity and strength (Gurung 1999).

Roles of Men and Women

Generally, as we described earlier, men and women do almost equal work, from land preparation to seed storage. Roles depend more on the family situation than on gender or ethnicity. Where a family has enough male members, women do not work much in the fields, whereas in families with fewer men, women worked equally side by side with them. However, ethnic sociocultural values are strong among the Brahmin-Chhetris; Brahmin-Chhetri women do not participate in ginger production as actively as Rai and Lepcha women.

Overall, women's work remains undervalued compared with men's. Women are seen, by both men and women, as helpers of men and, in general, their workload and is far greater than men's. Women not only work in the fields as much as or more than men, but they also have the major role in housework and caretaking.

Access and Limitations

By and large, women have almost equal access to all inputs required for ginger cultivation (seeds, land, labour). In all the families and households interviewed, both male and female members have equal opportunity to work in the fields, as well as to collect, select and store ginger.

However, there was one limiting factor for women: the Hindu belief in their 'impurity'. Menstruating Rai and Brahmin-Chhetri women are not allowed to enter ginger fields or even touch ginger seeds for 12 days.

Women have equal access to income from the ginger. However, this access clearly depends on their position in the household (for example daughter, mother, newly married daughter-in-law, mother-in-law). Normally, the more senior the woman in the household the more access she has to the money. Access does not necessarily mean that women have the freedom to spend cash income any way they desire. In all cases they first have to consult the family, especially the husband, and reach an agreement. Men, on the other hand, spend any money they receive to purchase whatever way they want (drinking, gambling, buying what they want) without consulting the family or getting approval.

A major limitation that women face concerns wages for agricultural labour. The wages of women labourers are systematically lower than those of men. This difference is based on the belief that women have less physical strength.

The market is considered a male domain. Women are considered to be less knowledgeable about the market. Although this was the case in all the communities we studied, it was more evident in the Brahmin-Chhetri community. However, changes are occurring. In all the communities the current generation of women is more actively involved in marketing than their mothers and grandmothers. More and more women are engaging in commercialization, either alone or along with a male member or members of their family. This trend also varies by type of household and among individuals.

Decision Making and Control

Decision-making opportunities and roles of women differ, both within households and within communities. Decision making is a complex process and is influenced by many factors. Women's input ranges from advice to recommendations and, in few cases, to actual decisions depending on her position in the family and household, and, to a certain extent, the community. Among the Lepcha and Rai communities, mothers and older daughters who are knowledgeable participate equally in selecting the site for ginger cultivation, whereas among the Brahmin-Chhetri farmers women are less involved in this task. Only female heads of household make this decision alone. The decision to buy and sell ginger is mostly made by the male head of the family, with women playing only an advisory role. However, differences exist depending on the family, position of the women in the household and individual people involved. Although in general women have a minor role in decision making, there are a few examples where women are equally, if not more, involved in this sphere.

By and large, women do not have much control over important assets like land. In most cases land is inherited by sons, leaving women without this important source of collateral for obtaining credit from formal institutions. Although equal inheritance rights are given to sons and daughters by the law under the 1956 Hindu Succession Act, custom still views men as the breadwinner and head of the family. Women do not seem to have much control over their own labour, as their wages are based on the male-dominated perception of women not being able to do as much work as men. Women's control of the money earned from ginger sales depends on their position in the household. The higher her position, the more control she has over this money, although the control is never complete because she must often spend it not how she pleases, but rather in response to the demands of her husband and family.

Image and Self-esteem

Our study shows that strong elements and symbols of cultural ideology explicitly or implicitly devalue women among the Rais and the Brahmin-Chhetri communities. Women are considered impure and unclean because of their childbearing role. These communities consider menstruating women 'impure' and do not allow them to enter agricultural fields. Similarly, after giving birth a woman is considered 'impure' for at least seven days; during this period the whole household is considered 'unclean' and no one is allowed to work in the fields. Women's image in these communities is one of inferiority. This is reinforced by the sociocultural taboo against women performing ginger rituals among the Rais, which has been adopted by the Lepcha and Brahmin-Chhetri as well as by other communities.

The self-image of women varies considerably. In general, younger women tend to have a better self-image than older ones. Young women have more self-confidence because of their greater exposure to the outside world and greater mobility. Most of the younger women think they are as good as men in every sense, including marketing of ginger. Older women tend to support and justify the existing male domination and their subservient roles as part of their cultural tradition, while the younger women are more aware of their rights and the inequity of their status. Their self-esteem varies from family to family, and ethnicity seems less a factor. In all cases, the situation at the household and family level has a major influence on women's image and self-esteem.

In summary, our analysis of gender relations in the region provides a complex web of confusing and sometimes contradictory evidence. There is evidence of the almost universal subjugation of women, but there are also signs that more equitable relations between men and women are emerging. Gender relations are not uniform in all households of a community and they differ between communities. They vary depending on individual men and women, and how they react and behave in various situations. Rai, Lepcha and Bhotia women have more autonomy, freedom of movement and opportunities to engage in entrepreneurial activities, assert their opinions, and influence important household and community decisions. At the same time, younger women in all the communities are taking a much more active role; they participate not only in the work but also in decision making and marketing. These changes are shaped by such factors as education, exposure to modern ideas, greater mobility, increased awareness of the outside world and increased political awareness. We found that younger women have much more confidence, a better self-image and more self-esteem than the older generation.

Some Key Issues and Challenges

The future of ginger production and commercialization is characterized by both constraints and opportunities. We conclude by describing some of those faced by women and by poor farmers.

The major constraint women face is the lack of ownership of land; in all cases it is the male head of the family who legally owns the land. This leads to another major problem for women: they have limited access to credit, especially from formal institutions like banks, as land is the most common collateral. Most women are overburdened with work as they undertake all household chores, shoulder the family responsibility of bringing food to the table, look after livestock, and work alongside men in the cultivation of crops. Most women have no time to take an active part in the marketing of ginger. Consequently, this limits their opportunities to control cash income. Socioculturally, it is not acceptable for women to be active and 'loud', especially in regard to marketing or other 'outside' activities; thus, they cannot take an active role in decision making and control.

In terms of opportunities (younger) Rai and Brahmin-Chhetri women especially are exploring new avenues. Many of them are self-confident and have high self-esteem. Rai women are demonstrating entrepreneurship, while Brahmin-Chhetri women have opportunities to participate in politics, particularly at the village level. They have become more aware of their rights. Many younger women are very much aware of things happening at the national and international levels. Rai women at Kharka-Sangsay said that they believed that the price of the ginger depends on international demand and markets, especially in Arab countries where demand for ginger is high. They expressed worry about the recent Iraq war as this might have a negative impact on the price of ginger.

Poor farmers have a limited amount of land and cannot cultivate much ginger as they require all available land to grow subsistence crops for their families. They also cannot afford to conduct experiments like those with more land. In Central Pandam poor farmers are usually migrants from Nepal and do not own any land legally. They cultivate land on a sharecropping basis or by leasing it. In addition, they do not have a right or access to the many facilities given to the farmers of that area by the state government. Poor farmers often learn cultivation and disease management practices from the farmers for whom they work as wage labourers, and this often helps them. Future action research could focus on ways to support them better.

Notes

1. Before 1706, Kalimpong was part of Sikkim, but after Bhutan invaded the eastern part of Sikkim that was contiguous to Bhutan and included modern-day Kalimpong, the adjoining area was heavily colonized and later completely annexed (Sen 1989). In 1865 British India annexed Kalimpong from Bhutan following the Indo-Bhutan war of 1864 and included it in Darjeeling district in 1866 (O'Malley 1985; Subba 1992).
2. Kalimpong had been the major ginger growing area of the region, but after the annexation of Sikkim by India in 1975, the country's ginger market opened for Sikkim as well. From then on, the region as a whole (Kalimpong and Sikkim) became one of the major ginger suppliers for the Indian market (15 per cent of total production).
3. In this process, called *jar phook*, the medicine man touches the sick person from head to foot with a piece of ginger while chanting. The medicine man then chews the ginger and applies it to the affected area. This ritual is done early in the morning before either the medicine man or the sick person has eaten anything.
4. *Parma* is a system of traditional labour exchange, whereby one member of a family or household goes to help or work for another household. The exchanges are rotational.
5. Approximate conversion: 1 United States dollar (USD) = 43.6 Indian rupees (INR).

References

Basnet, Lal Bahadur (1974). *Sikkim: A Short Political History*. New Delhi: S. Chand and Co.
Chemjong, I.S. (1966). *History of the Kirat People: Parts I and II*. Kathmandu: Aathrai Publications.
Fonning, A.R. (1987). *Lepcha: My Vanishing Tribe*. New Delhi: Sterling Publishers.
Government of Sikkim (1993). *Sikkim: A Statistical Profile, 1979–80 to 1991–92*. Gangtok: Bureau of Economics and Statistics, Department of Planning, Government of Sikkim.
Gurung, J.D., ed. (1999). *Searching for Women's Voices in the Hindu-Kush Himalayas*. Kathmandu: International Centre for Integrated Mountain Development.
O'Malley, L.S.S. (1985 [1907]). *Bengal District Gazetteers: Darjeeling* (2nd ed.). Calcutta: Bengal Secretariat Book Depot; Delhi: Lagos Press.
Pinn, F. (1986). *Road to Destiny: Darjeeling Letters 1839*. Calcutta: Oxford University Press.
Rao, Raghunandha (1978). *Sikkim: The Story of its Integration with India*. New Delhi: Cosmo Publications.
Sen, Jahar (1989). *Darjeeling: A Favoured Retreat*. New Delhi: Indus Publishing.
Subba, Tanka B. (1992). *Ethnicity, State and development: A Case Study of Gorkhaland Movement in Darjeeling*. New Delhi: Har-Anand Publications.
Watkins, J. (1996). *Spirited Women: Gender, Religion and Cultural Identity in the Nepal Himalaya*. New York: Columbia University Press.

3

Strengthening Market Linkages for Women Vegetable Vendors

*Experiences from
Kohima, Nagaland, India*

Photo credit: NEPED

VENGOTA NAKRO AND CHOZHULE KIKHI

Geography and Issues

The Indian subcontinent is separated from the rest of Asia by a continuous barrier of mountain ranges. The country is densely populated, with over 1 billion people—nearly one-sixth of the world's population. It is a country with enormously varied cultures, landscapes and history. North-east India is made up of seven states (Arunachal Pradesh, Assam, Manipur, Meghalaya, Mizoram, Nagaland and Tripura) known as the 'Seven Sisters'. They are linked to the rest of India by only a narrow strip of land. This relative isolation lends a distinctive quality to the lifestyles and cultures of the area, which are still predominantly tribal. Remoteness and inaccessibility make the north-east one of the most sparsely populated regions in the country. Arunachal Pradesh has the lowest density, with only 13 people per km^2. The tribal people, who constitute the majority, are mostly of Mongoloid ancestry and originated from China, Tibet, Thailand and Myanmar. Nagaland is inhabited by 17 groups of the Tibeto-Burmese tribes who were once 'headhunters'. North-east India is known as a 'hot spot' for biodiversity of natural resources. This region also has a sensitive border zone where India meets Bhutan, China, Myanmar and Bangladesh.

Modern industries are lacking. Traditional industries are based on handicrafts. Handloom weaving is the major cottage industry in all the hill states, and women are the principal weavers. The forest provides timber and bamboo. Agriculture in the north-east is divided between settled farming in the rice-growing valleys and *jhum* or shifting cultivation in the hills. More than 70 per cent of the people live off the land and grow rice and maize as their main crops. The region is one of the wettest in the world, yet in many hill towns there is an acute shortage of drinking water as a result of indiscriminate felling of trees. There is potential for hydroelectric power, but it has not been developed yet.

Nagaland has an area of 16, 579 km^2 and borders Myanmar in the east (see Figure 3.1). The climate ranges from subtropical to subtemperate with altitudes ranging from 200 to 3,800 m above sea level. Annual rainfall is 2,500 mm. Because of these geophysical characteristics, farmers can cultivate subtropical and subtemperate crops simultaneously

Figure 3.1
The Nagaland Region Showing the Location of the Research Sites

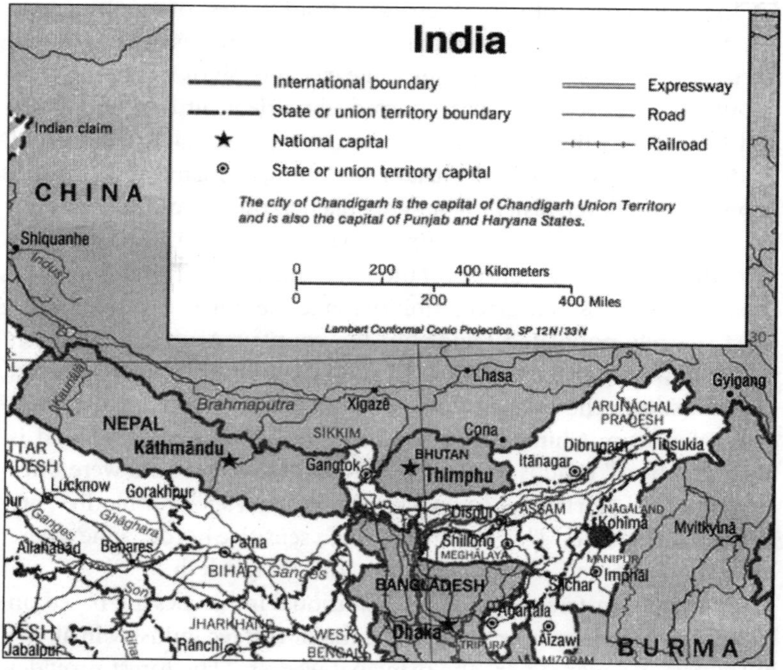

Source: Perry-Castañeda Library Map Collection.

and in the same location with comparative ease. They are able to produce crops under natural conditions: no external inputs, such as chemical fertilizers, are needed.

Seventeen major tribes reside in more than 1,000 villages. Villages are perched at altitudes above 100 m, and some are in the extreme highlands at over 2,000 m. Nagaland is considered to be the northernmost outpost of the rainforest, providing habitat for all its typical families, especially the Dipterocarpaceae. This habitat is also home to numerous wild and cultivated vegetables. Agriculture, primarily slash and burn, occurs on moderate to very steep slopes. Some tribes, particularly the Angami and Chakhesang, practise terrace rice cultivation on steep slopes.

There is immense potential for vegetable production at the village level to feed the growing population in the townships. Demand exists and farmers have the knowledge to produce a large quantity, good quality and wide variety of produce.

Over 70 per cent of the inhabitants of Nagaland practise *jhum* cultivation, primarily because the terrain makes terracing and irrigation difficult. The *jhum* system is well suited to the subsistence farmers living in the hills where rice may be intercropped with no fewer than 20 to 60 other species in one field. There is an intimate link between *jhum* cultivation and crop biodiversity. For example, in the past, 30 types of millet were cultivated in Viswema village; now several varieties have been lost due to the change from *jhum* to terrace cultivation. Normally, after two to three years of use, fields are left fallow to allow the soil to rejuvenate for five to 20 years, depending on the availability of land.

Unfortunately, forest and soil degradation are major problems. According to the Department of Forest and Ecology roughly 33 per cent or 2,843 km^2 of the total 8,629 km^2 of classified forests were degraded in 1996–97. Of the total forested area, about 90 per cent is privately owned (this includes village ownership), while only about 10 per cent is owned by the state. In addition, with a growth rate of 6.4 per cent annually, population pressure and scarcity of land threaten traditional food security (Government of Nagaland 2001).

Women are considered to be the primary gatherers of forest resources to meet household needs; they collect wild vegetables (according to informants from one village, more than 50 varieties), non-timber forest products and wood for fuel. Depletion of these resources, therefore, has a direct impact on women, likely increasing their workload and drudgery. It also has a direct impact on the overall livelihood of the people who depend on forest resources not only for their own use but also as a crucial source of family income.

Sustainable development in Nagaland depends primarily on a balanced approach that includes biodiversity conservation, sustainable management of existing lands and forests, the exploitation of new forest resources, and adapting agricultural systems over time. Expansion of the agricultural area into accessible primary forest continues to be a concern.

Research Objectives, Questions and Methods

In this context, we set out to support women vegetable vendors seeking to increase their income and benefits through better understanding of enabling factors and constraints in the local vegetable markets. The main research questions were:

1. Who does what in vegetable production or collection and sale?
2. What are the perceptions of the various social groups of women vegetable vendors?
3. What is the impact of vegetable selling by women on the family and the community?
4. What are the enabling factors and constraints on women who sell vegetables?
5. How can women overcome the constraints and build on the enabling factors?

We wanted to learn more about the lives of women vegetable vendors and share this knowledge with others involved in or related to this activity. We were also hoping that the vendors would make use of the knowledge to increase their income and other benefits. As researchers, we were looking for ways to play a supportive role and also provide technical inputs to the women. We also intended to enhance the knowledge of all participants about social and gender analysis and make them more sensitive to gender issues in particular. The research contributed directly to the Nagaland Empowerment of People through Economic Development (NEPED) project (see Box 3.1).

The research was carried out by a small team in collaboration with NEPED's Project Operations Unit (POU), women vegetable vendors and their families, the village council, the village development board, community representatives, members of the Kohima Town Committee and chamber of commerce, and the district agriculture and horticulture officers from Kohima.

We used a variety of tools, including questionnaires for both semi-structured interviews and group discussions. We carried out a preliminary systematic sample survey in Kohima, the capital of Nagaland, to

> **Box 3.1 The NEPED Project**
>
> The NEPED project is jointly funded by the India–Canada Environment Facility, and the Government of Nagaland. Its purpose is to provide a mechanism for sustainable community-based land and natural resource management to reduce poverty and increase livelihood options in Nagaland. The main activity is to establish a sustainable revolving credit scheme in 104 selected villages in eight districts of Nagaland.
>
> The project team or Project Operations Unit (POU) consists of 13 members including the team leader, who is the agriculture production commissioner of Nagaland. Other members are officers selected from state government departments. All the POU members work in the project as equals, irrespective of their position in the parent department.

identify villages where women vegetable vendors come from, the number and sources of vegetable species being sold, and the ways in which the vegetables are sold. We employed one field person for this work. From the list of villages, we chose vendors from four villages based on criteria such as accessibility, distance from Kohima and willingness to take part in the research as key informants. The villages and their distances from Kohima were: Khonoma and Tsiese Basa (22 and 18 km, or a moderate distance away), Merema (12 km) and Pholami in Phek district (107 km). The villages can be reached by all-weather roads. They are covered by the NEPED project and, as such, we had already established a rapport with the respondents.

We first visited the selected villages to discuss the purpose of the research with village functionaries and to obtain their consent. Through these discussions we identified women vegetable vendors who would be willing to take part in the research. The vendors were encouraged to form self-help groups and some training was provided. One group member from each village was chosen as 'point person'—the person whom project field staff would contact. These people were given hands-on training in basic research and communication methods.

We carried out semi-structured interviews and held informal discussions with government officials and Kohima Town Committee members. We also encouraged the formation of a consultative committee at this decision-making level to explore avenues for using the research results over the long term. We used several participatory rural assessment (PRA) tools:

1. a seasonal calendar of crops to help us understand when the various vegetables fetch the best price and when various crops glut the market;
2. semi-structured interviews to gain an understanding of the perceptions of producers, vendors and the community regarding how to increase family income and the associated constraints; and
3. a survey of 20 households in the villages of Merema and Tsiese Basa to find out in more detail about the crops being sold, key constraints, production costs, grading and dressing, and the practice of storing unsold vegetables for the next day. The survey was also expanded to 63 households to determine the total cash income from the sale of vegetables during the last three months.

In addition, we used focus group discussions to validate results of other forms of data collection, to elicit group perceptions about the marketing of various crops, and the community's perceptions of enabling factors and constraints. Opportunities and gaps in vegetable production and marketing were also discussed with village functionaries. We interviewed key informants to validate information collected, draw out 'best management practices' that could be emulated by other producers and to learn about strategies to increase production and marketing. We did a transect walk to observe the production of vegetables (noting such features as distance from homes, size of the fields and crops grown). We also made direct observations of collecting, sorting, grading and selling processes.

Most of the regular women vegetable vendors are from Kohima and are descendents of the original settlers of the village. To facilitate group discussions, we contracted a local person, who was able to dissipate some of the vendors' fears and suspicions. Building rapport took less time than expected; however, the vendors were often hesitant to provide complete information. We sensed that they were afraid that the information collected would be used against them. Although we assured them confidentiality of our research findings, we did not manage to validate everything. It is also important to note that most vendors who participated in the study are poor, which may have led to an underestimation of enabling factors and overstatement of constraints. We also had to face unrealistic expectations by some participants in terms of what they thought to gain from the research(ers).

Vegetable Production Methods

Farmers have two main sources of vegetables for household consumption and sale: wild and cultivated. Wild vegetables are collected from forests, firewood reserves and *jhum* fallows. Generally, there have been no restrictions on collection; however, due to overexploitation of these species in recent years, officials in some villages are now imposing a ban on their collection for sale. According to the women vendors, these restrictions are not ecologically sound for some plants, especially the Balsamaceae family. According to the women, these species thrive better when the new shoots are nipped, as this results in branching and spreading. Left alone, they would grow into tall bushes, then die.

Most of the wild vegetables can be easily propagated through seeds and vegetative cuttings, provided the growing conditions in the domestic site are similar to those in the wild. Such sites are available in pockets. Further action research could include a domestication trial using *Gajo jotho*, a popular vegetable of the Balsamaceae family.

The farmers in the study villages practise an agriculture system that consists of three components: terraced rice cultivation (TRC), *jhum* and home gardens. Each contributes to the supply of vegetables for sale at a particular time, and together they ensure a constant supply of fresh vegetables in the Kohima market.

Terraced Rice Cultivation

Although this type of cultivation is used mainly for rice, in recent years relay cropping with vegetable cash crops is becoming more common. For example, tomato seeds are scattered in the rice fields during

September and October when rice plants are flowering. The seeds float in the standing water, but as the water level recedes as harvest time approaches, the tomato seeds come into contact with the soil, well soaked and ready to germinate. By late May they are ready for harvest and continue to be sold in the market until July. The tomatoes produced this way are usually cherry tomatoes; they have a very short shelf-life and do not fetch a good price, although they taste good and are organically produced. Other relay crops include rice beans, grown in the terrace risers, and banana, brinjal, chilli and other assorted vegetables and fruits grown on the margins of terraced fields.

An action point in this area might be to increase the income of producers and vendors by introducing varieties that have a longer shelf-life and develop packaging techniques that increase shelf-life, as a significant proportion of the crop is lost during transport because it is delicate.

Jhum (Shifting Cultivation) Fields

Subsidiary crops are grown in *jhum* fields. In the past the main crops were maize, jobstear, millets and other cereals. Lately, there has been a significant shift, with cereal crops being replaced by vegetable cash crops, bulky vegetables like pumpkin being replaced by cucumber, and perilla being replaced by *Solanum* spp. In general, vegetables with a longer shelf-life, higher price and smaller volume are replacing conventional crops with lower economic returns.

Home Gardens

Most vegetables for home consumption and for sale are grown in home gardens, which are usually near the homestead or on the outskirts of the village. Because they are closer to home than the other components of the system, they receive the best care. In these home gardens a mixture of trees, perennials, creepers, climbers and annual crops are cultivated. More than 120 species have been reported growing in Tanhai village in Mon district.

THE KOHIMA VEGETABLE MARKET

Kohima is the capital of Nagaland and has a population of approximately 83,000. It includes people from all the Naga tribes and other Indian communities. The various groups use many different plants and animals in their regular diets and most are available in Kohima markets.

We observed that 90 to 95 per cent of the vendors are women. Women look after the management, cultivation, harvesting, and processing of the crops, although men sometimes help bring the produce to the local market. Although some goods are sold in bulk to regular vendors, others are sold directly to consumers on a temporary site. The four types of vegetable marketing enterprises operating in Kohima are shown in Table 3.1. As a result of their role in trade, women achieve a certain control over household decision making. Most of the income is used to buy essential commodities.

Table 3.1
An Overview of Vegetable Vending Enterprises in Kohima

Type of enterprise	No. of entrepreneurs	No. of clusters	No. of vegetable species marketed	Source of vegetables	Entrepreneurs
Naga vegetable vendors			59	Villages	
Regular	136	5			Women
Part-time	207	12			Women
Wholesale dealers	12		19	Dimapur and outside Nagaland	Family
Vegetable retail shops	120	10	19	Dimapur and outside Nagaland	Family
Home delivery vendors	50	12	19	Dimapur and outside Nagaland	Men

Source: NEPED fieldwork.

Naga Vegetable Vendors

This group dominates the market; the approximately 340 vendors are clustered in 12 locations, and among them they sell about 60 vegetable species. The produce is mostly organic and comes from neighbouring villages as well as some more than 100 km away. Most of these vendors are producers themselves. They can be divided into two subgroups: regular and part-time vendors. Regular vendors have a permanent space in the market, under a shed roof constructed by the authorities, for which they pay rent. Selling vegetables is their profession. They buy produce at wholesale rates from temporary and part-time vendors early in the morning. Part-time vendors, on the other hand, are occasional sellers. They bring their produce to market when there is a surplus from the field or home gardens or when they cultivate a special crop to sell.

Vendors bring their produce to market on the community bus or by carrying it on their head, depending on the distance and quantity. Regular vendors often make bulk purchases from farmers, but in many cases the farmers choose to sit at a vantage point on the roadside and sell their own vegetables. A kind of love–hate relationship can development between regular and part-time vendors. The former depend largely on the latter to supply them with produce for sale. But when the part-time vendors see that the regular vendors are able to buy their vegetables at a competitive price and sell them at a good profit, they often decide to retail their produce themselves. Regular vendors see the squatters as a threat because they compete for customers, and have asked the Kohima Town Committee to evict them. Often both parties lose and ill feelings result.

Wholesale Dealers

Wholesalers mostly sell vegetables brought in from the foothill plains of Dimapur and Assam by trucks travelling regularly between Kohima and Dimapur. Most of the vegetables they sell have a long shelf-life. They bring in goods in bulk and supply vegetable retail sale shops and home delivery vendors. During the potato and tomato season they sometimes buy from villagers around Kohima and sell these crops to retailers. About 12 wholesalers cater to 120 vegetable retail shops and about 50 home delivery vendors.

Vegetable Retail Shops

The 120 shops in Kohima sell 19 different vegetables at 10 locations. Most of these vegetables are bought from the wholesale dealers on a regular basis, but small quantities are sometimes purchased from the part-timers to supplement the variety of produce for sale in the shops.

Home Delivery Vendors

These vendors buy vegetables from the wholesale dealers and carry them from house to house. They can carry 30 to 40 kg of vegetables on their head.

Market Regulation: The Efforts of the Kohima Town Committee

To help part-time vendors sell their vegetables at a profit, the Kohima Town Committee provides infrastructure at three locations and issues directives that the vendors must adhere to. To make things easier for pedestrians and to avoid traffic congestion, vendors coming from different localities are assigned to different locations (Table 3.2).

Only the Super Market and Mao Market locations are being put to full use; Old Ministers Hill (Jail Colony) is partly used and Keziekie and the TCP gate are used for a weekly market. The reasons part-time vendors give for not using the infrastructure are plain and simple: there are no customers there. A lesson can be learned from the initiation of Mao Market, which was started only about five years ago by some elderly people from the Mao community. As the market was for the benefit of one community, the vendors from that community crowded the place to sell their vegetables, and it quickly became a popular marketplace.

The Kohima Town Committee members sympathize with the women vegetable vendors, but they are also concerned about keeping the town in order. They are aware that the rural women vendors squatting at the roadside are poor and may have invested all their savings to buy vegetables in their village to sell in Kohima. From time to time the committee members attempt to enforce the rules by evicting the wayside vendors, but this is only temporary. The vendors simply move to another part

Table 3.2
Designated Market Niches for Vegetables Vendors Provided by the Kohima Town Committee

Source of vegetables	Designated market	Status of vendors	Remarks
Phek district Northern I and II Wokha Tseminyu	Keziekie daily market	Part-time	Free of tax
Southern Angami	Old Ministers' Hill market (Jail Colony)	Part-time and some regular	Nominal tax or free
Western Angami Jalukie Chakhro (Dimapur)			
Jalukie Western Chakhro (Dimapur)	TCP gate, Kohima	Part-time	Free of tax
Regular retailers	Super Market	Regular	Tax to town committee
Direct producer	Mao Market, IOC	Regular	Privately owned

Source: Kohima Town Committee.

of town and continue to do their business as usual. Shopkeepers, especially those from the mainland, are unhappy with these vendors because they obstruct the entrance to the their shops. Local shopkeepers have more respect for the Naga women and are less likely to assert their rights.

The Kohima Town Committee introduced weekly markets in various locations to woo the part-time vendors. Information about the weekly market day was circulated to all neighbouring villages. However, the committee has done nothing to determine whether the local vendors are benefiting from this initiative. We noted that these locations are mostly used by vendors from the mainland.

Although all four types of vendors are playing important roles in the marketing of vegetables, our research and the following information concentrates on the Naga vegetable vendors (regular and part-time) in the four chosen villages: Pholami, Merema, Tsiese Basa and Khonoma.

Pholami

This village is located in Phek district about 107 km south-east of Kohima, and its inhabitants are Chakhesang. There are 274 households with a population of about 1,250. The road through the village is usable in all seasons, enabling a community bus to travel between Pholami and Pfutsero, the nearest town. In this village there are 34 women vendors who sell vegetables regularly in the four neighbouring towns of Phek, Chizami, Pfutsero and Kohima. These vendors fall into three categories:

1. Those who grow all their produce in their own fields and gardens (15 vendors). These vendors are the main suppliers to the regular vegetable vendors in town. Because they grow the vegetables themselves and sell them in bulk, they are assured a reasonable price and both the trader and producer can profit.
2. Those who grow about half the vegetables they sell, procuring the rest from others (six vendors).
3. Those who purchase all their vegetables from producers (13 vendors).

During December 2003 and January 2004 the total cash income generated by the women from vegetable sales was about INR 110,000 (United States dollar [USD] = 43.6 Indian rupees [INR]). This is a significant amount, as there are virtually no off-farm employment opportunities in the village. The main regular source of cash income in Pholami is the salary of government employees, mainly primary school teachers living in the village.

Enabling Factors

In recent years there has been an increase in the number of women selling vegetables from the village. An elderly male village council member remarked: 'The wealth of the village can be measured by the quantity of vegetables being taken out from the village for sale. The more the quantity, the richer the village.' Several factors encourage the vendors to sell vegetables outside the village. The desire to provide for their children's education and to improve housing and living conditions, among other reasons, have prompted rural farmers to look for income-generating opportunities. Growing and selling vegetables is a way to generate cash in the shortest time.

All vegetable vendors from this village stated that they are able to sell all their produce. However, they also said that at times there is a glut in the market for certain vegetables, which may then have to be sold at throwaway prices.

Factors that enable the farmers to undertake commercial production include favourable growing conditions, support from their families and the community, availability of transportation to market, and avenues to generate income and other benefits.

Favourable Conditions for Vegetable Production. To estimate the number of varieties of vegetables and fruits produced in the village for sale, a focus group of women vendors of Pholami village was assembled. Table 3.3 indicates that in January, February and March the number of species is lowest, whereas from September to November is highest. The total number of fruit and vegetable species grown for sale in this village is about 50.

Table 3.3
Number of Vegetable Species Sold Each Month by Women Vendors from Pholami

Month	No. of wild vegetables	No. of cultivated vegetables	Total no. of vegetables
January	8	6	14
February	8	6	14
March	8	6	14
April	11	14	25
May	11	14	24
June	11	19	30
July	11	22	33
August	11	22	33
September	11	28	39
October	11	30	41
November	11	27	38
December	11	19	30

Source: Group discussion with women vendors from Pholami.

The 10 most common wild vegetables are three species of Balsamaceae, three species of ferns, wild pepper leaves, two species of *Oenethe* and *Centella asiatica*. Some of these (Balsamaceae and wild pepper leaves) are collected mainly from primary forests; others, such as ferns, are found in secondary forests. *Centella asiatica* is collected from the rice terraces. There are no restrictions on the collection of wild vegetables even from private land; on the other hand, cultivated vegetables and fruits cannot be taken from someone else's field, although they can be eaten in the field.

Many of the wild vegetables are available all year, especially the most common ones (see Box 3.2). Several cultivated vegetable are also sold throughout the year, especially those that can be consumed wholly.

According to the vendors, men and women collect the wild vegetables jointly. On almost every trip to the market, the vendors bring 50 to 60 bundles of wild vegetables; they have noticed no decrease in abundance of these species.

Farmers cultivate different crops at different elevations depending on site conditions. During a transect walk, we observed that vegetables such as cabbage, field peas and potatoes are cultivated in the higher

---

Box 3.2 *Gajo-jotho*: A Promising Wild Vegetable for Domestication

Of the wild vegetables, *Gajo-jotho*, a species in the Balsamaceae family, is the most favoured because:

- It is abundant in the wild and easy to collect in large quantities in a short time.
- It has a long shelf-life. The vendors told us that although it wilts on exposure to sunlight, it regains its freshness after they soak the stems in water overnight.

A possible intervention is to domesticate this crop in locations such as home gardens under conditions similar to those found in the wild. The vendors observed that rooting takes place at nodes that come into contact with soil. They think that it could be propagated easily.

reaches, whereas ginger, garlic, chilli, beans and taro are cultivated at lower altitudes. Most of the fields are on the outskirts of the village where villagers maintain a kind of sedentary *jhum* field.

Support from the Family and Community. Vegetables are harvested a day before the vendors travel to the market. Grading, sorting and cleaning is shared by family members the same evening, making it possible for the vendor to board the bus very early the next morning. Family members also help carry the goods to the bus stand and load.

The community supports vegetable entrepreneurship as well. For example, the community has installed a strong fencing system to keep cattle out of agricultural areas. The system is a great help in terms of agriculture production of winter crops (see Box 3.3).

Pholami community members who are permanent residents of Kohima support women vegetable vendors by providing them with shelter overnight. Thus, the vendors have free lodging and food, making it possible for them to take home all their income from sales. They also receive guidance from these residents. The women vendors receive support and encouragement from the family while they are away; during their absence, for example, the men take care of the household chores.

Box 3.3

Fencing: A Leap towards Sedentary Agriculture

Traditionally, cattle were released and allowed to graze anywhere from the second week of November to the end of February. As a result, farmers were unable to cultivate winter or perennial crops. To encourage cash crop production, the Pholami village council resolved that all cattle would be confined to a demarcated area within the borders of the village. This rule was enforced to the letter and in spirit. Fencing has allowed a significant area to be cultivated for cash crops and many farmers are benefiting from the increased harvest. There has also been a significant increase in the number of women vegetable vendors.

Availability of Transportation. A private bus operates regularly between Pholami and the town of Pfutsero, leaving Pholami at about 5 A.M. The bus owner told us that the vendors are often the only people travelling to Pfutsero. During the peak vegetable harvest, competition for space on the bus is so stiff that quarrels among the vendors become common and the bus is often overloaded. Efforts are being made to schedule vendors' travel throughout the week to avoid these problems, although this is complicated because each vendor has her own activity schedule.

Avenues for Generating Income and other Benefits from Vending Vegetables. With the population in the township expanding due to natural growth and rural-to-urban migration, the demand for fresh vegetables is on the rise. Farmers in the villages are responding by shifting their crops to vegetables that have a higher market value. For example, in the past, ginger was cultivated for home consumption only. Now the ginger fields are expanding so that some of the crop can be sold. (For more about ginger production, see Chapter 2.) Production of mustard leaves, cabbages and field peas is also increasing.

Vendors' purpose is either to generate cash income by selling their own produce or to make a profit in the case of those who retail. Both strategies are relatively successful due to the growing demand for organically produced vegetables in Kohima and other markets (see Box 3.4).

Box 3.4

Insect Infestation: A Selling Point!

Two years ago one of the progressive farmer of Pholami and his family started to cultivate cabbages only, instead of shifting among the over 20 different crops. The cabbages were generally healthy, except for an infestation of insects on the outer leaves. Consequently, the family had trouble selling the crop. They hired a truck and brought the cabbages to Phek, the district headquarters. Here the cabbages sold 'like hot cakes'. The selling point was the insect infestation—according to one of the buyers, it was irrefutable proof that the product was organic!

To learn more about how vendors are generating income, we did two contrasting small case studies focusing on the costs and benefits of part-time vegetable vendors.

Because most farmers in the villages practise a mixed cropping system, it is difficult to devise an acceptable way to quantify production costs per unit area. In a *jhum* field a farmer can cultivate as many as 60 species (Supong 1999). Average *jhum* fields in the study villages contain about 20 different crops. We were not able to segregate the labour component for each crop, nor for the quantities harvested. The farmers follow a complex cycle of mixed cropping, relay cropping, multiple-cropping and perennial cropping. As an alternative, we estimated the costs and benefits of selling vegetables for two part-time vendors.

Vendor 1 (part-time, few resources)   This vendor usually travels to Kohima once a month to sell vegetables. In 2003 she was able to do this only twice as she was nursing a child. She has a relative in Kohima, where she can stay for the night at no cost (there are no public overnight shelters in town). It takes five days to sell her produce. Table 3.4 presents the costs and benefits for one regular trip.

From these numbers it is clear that if the vendor did not have a relative with whom to spend the night in Kohima, she would not be able to generate much profit. This raises the question of whether she would be better off selling her produce to a wholesaler in Kohima. She would save a day in labour costs, but her income would diminish by approximately one third, which explains why vendors prefer to retail the vegetables themselves.

Vendor 2 (part-time, more resources)   In contrast with vendor 1, this vendor from the same village has more resources to invest and is making a substantial profit in her venture. She has seven years of vending experience and has learned that specialization is most profitable. Instead of spending time on many items as is usually done, she selects only products and invests as follows (Table 3.5).

Compared with vendor 1, we can see that experience, contacts and the careful selection of and concentration on a few items may make for a more profitable venture. Given that the number of days involved in

Table 3.4
Costs and Benefits of a Trip to the Kohima Market for Vendor 1 (INR)

Item	Expenditure	Income
Produce		
Tree tomato, purchased (about 600)	170	550
Bananas, purchased (about 350)	100	350
Bananas, own garden	120	250
Balsamaceae species, collected in 1 day	70	370
Colocassia shoots, purchased	50	100
Ginger inflorescence, purchased	50	120
Subtotal for produce	560	1,740
Transport and sundries		
Bus fare from village to the nearest bus stop	40	
Bus fare from the nearest bus stop to Kohima	100	
Loading charges	20	
Taxi fare	50	
Lunches while travelling (2)	50	
Return bus fare	100	
Subtotal for transport and sundries	360	
Other costs		
1 day's wage for collecting wild vegetables (man)	70	
1 day's wages for collecting vegetables (woman)	60	
2 days' wages for journey (to and from market)	120	
2 days' wages for selling vegetables	120	
Subtotal for other costs	370	
Grand total	1,290	1,740
Profit		450

Source: NEPED fieldwork.

preparation, travelling and selling is the same, it is more lucrative to acquire a larger quantity of goods. The vendors report that in most cases the producers will give them vegetables on credit; thus, they do not need to have cash in hand except for transportation. We can also see that marketing bananas is more profitable than selling other produce.

Women vendors squat for long days at the roadside, defying all obstacles such as traffic jams, the ire of the KTC, rain, wind, heat and other discomforts to sell their vegetables. We asked them why they do not sell their vegetables to a wholesaler and return home; on the surface,

Table 3.5
Costs and Benefits of a Trip to the Kohima Market for Vendor 2 (INR)

Item	Expenditure (INR)	Income
Produce		
Bananas	1,250	4,100
Tree tomatoes	100	200
Colocassia shoots	50	100
Subtotal for produce	1,400	4,400
Other costs		
Transportation	385	
5 days' wages (INR 60/day)	300	
Subtotal for other costs	685	
Grand total	2,085	4,400
Profit		2,315

Source: NEPED fieldwork.

this option seems better, allowing both parties to benefit. We discovered that vendors can make three times as much selling produce themselves than they would by selling it to a wholesaler (Table 3.6).

Table 3.6
Comparison of Retail and Wholesale Prices per Unit of Vegetables and Fruits Obtained by a Vendor from Pholami Village (INR)

| Produce item | Cost at source | Selling price at Kohima | |
		Wholesale	Retail
Bananas (each)	0.25	0.50	1.00
Tree tomatoes (each)	0.25–0.33	0.33	0.80
Ginger leaves (bundle)	5.00	10.00	10.00
Colocassia shoots (bundle)	4.10	6.70	10.00
Passion fruit (each)	0.17	0.33	0.66
Chayote (each)	1.00	2.00	3.30
Cucumber (each)	1.40	3.30	4.00
Local garlic (bundle)	10.00	15.00	30.00
Lablab (bundle)	5.00	10.00	10.00
Field peas (kg)	10.00	12.50	20.00

Source: NEPED fieldwork (group discussion with women vendors from Pholami).

Non-Material Benefits of Selling at Kohima. For some vendors, selling their produce in Kohima allows them keep in touch with their children, who are at school and college there. When the parents come to the market, they also visit their children and pay school fees. More than 340 Naga women vegetable vendors are going about their business every day in Kohima. Although at first most are strangers, they make friends and learn production and marketing techniques from each other. Some seasoned part-time vendors have built such good relations with regular vendors that they can dispose of their vegetables at a reasonable price at any time of the day. Many new part-time vendors have learned the art of sorting, packaging and quality control by selling vegetables themselves. Spending time in Kohima also allows the women to learn about other town business. They find out where to buy the best priced food and items that might not be available in Pfutsero.

Constraining Factors

Transportation Deficiencies. The single bus travelling between Pholami and Pfutsero is reliable but often insufficient to meet the demand. It best serves vendors who sell their vegetables at Pfutsero. However, more than 10 vendors sell their vegetables in Kohima and need to travel another 74 km, which means hitchhiking on whatever truck is available. On arrival in Kohima, they have to hire a taxi to transport the vegetable to the spot where they set up business. This journey takes a toll on the vendors. During March, April, May and December (the peak season), on Tuesdays, Wednesdays and Thursdays, there is a mad rush for the bus. Many vendors are unable to get a seat with the result that their produce goes to waste.

Insufficient Production for Wholesale Trading. Despite the favourable growing conditions described earlier, villagers have difficulty producing or procuring enough vegetables to allow them to venture into wholesale trading, which, according to the women vendors, would be worthwhile. There are several reasons for this difficulty. First, most villagers believe that vegetables are best produced in home gardens. Only a few progressive farmers have tried alternatives, such as growing cabbages and field peas in TRC fields as winter crops. Strong manpower is required to plough fields in November as the ground is heavy with moisture; this is not readily available as many men are busy with other work at this time. Second, continuous cultivation of crops in the same area leads to nutrient depletion and diseases, and farmers have no access to fertilizers and only limited knowledge of soil fertility improvement practices. Third, according to some women vegetable vendors, cultivation could be increased if the men cleared the jungle to create more farmland. They consider the onus for increased vegetable production to be on men, who are clearly failing to respond.

Disputes among Vendors. Currently, there are many vendors and competition is strong. According to villagers, the number is increasing every season. Key informants reported that some fellow villagers tell producers that the vendors are making a huge profit. This kind of hearsay has prompted several producers to sell their own crops, thus adding to the mêlée. As a result of the competition among vendors, producers have escalated their prices and vendors' profits have decreased.

Challenges Concerning the Sites of Vending Activity. The KTC has designated specific sites for vendors coming from villages in different directions. KTC members assume that vendors comply with these directives, but part-time vendors are not always aware of them or deliberately ignore them and set up wherever business is good. From time to time, the KTC raids the wayside vendors and seizes their goods. As Kohima's population is increasing, the vendors are finding it more and more difficult to find a place to sell their vegetables. The increasing number of vehicles is also a serious and growing problem.

Intervention Strategies

During the research process, we identified action-oriented strategies to help the vegetable vendors overcome some of the obstacles they face. To address the transportation bottleneck, we held discussions with village functionaries and community leaders to assess both current and potential vegetable production if this barrier could be removed. The community suggested that the problem could be solved by making a public carrier truck available in the village. One progressive farmer-entrepreneur showed interest in this idea, and we are exploring avenues by which the farmer could get a bank loan to purchase a truck.

To address the problem of insufficient supply, we carried out a thematic resource mapping exercise with both men and women in the village. They mapped out what kind of vegetables could be produced where. The participants were very knowledgeable about cultivation of indigenous crops and identified options for increasing production (see Box 3.5 for an example that has already been put into practice). In partnership with four progressive farming families, we provided vegetable seeds and labour for the cultivation of vegetables in TRC fields as a demonstration. The income generated is expected to have a ripple effect in the village.

The farmers know that settled cultivation is only possible if nutrients are added to the soil, but they have no access to fertilizers. To overcome this problem, we initiated a dialogue with a Bio-fertilizer Laboratory (of the Agriculture Department, Government of Nagaland) so that the research team could facilitate the supply of bio-fertilizers in the low quantities that are required.

Group discussions were initiated with the men to make them aware that they need to devote more time to cultivation of winter crops in November. With the researchers facilitating the discussion, we found it was effective to invite women vendors to speak about the benefits that could be accrued from this increased production.

We encouraged the establishment of a consultative committee to look into the problem of marketing sites in Kohima, drawing lessons

---

Box 3.5 — Substituting Vegetables for Rice in TRCs

In 2003 one of the progressives farmers of Sakraba village grew vegetables in one of his TRC fields instead of rice. He was able to generate much higher income than he would have by growing rice in that field. As a result, he has leased a TRC field from a friend by paying him the full amount of rice that the field yields. He is expecting to make an even larger profit than he did in 2003.

> **Box 3.6 — Creating an Overnight Storeroom in Kohima**
>
> Busuveyi, a Pholami villager, is a resident of Kohima. He understood the benefits that women vegetable vendors are generating for their community and offered to partition a room in his business premises for overnight storage of unsold vegetables. We helped him carry out this plan.

from the experience and the success of the Mao Market consultative committee. We also assisted with the creation of an overnight storage area for vendors selling on consecutive days in town (see Box 3.6).

Merema and Tsiese Basa

These two villages are north of Kohima: Merema is 12 km away and Tsiese Basa is 18 km from Kohima. The types of vegetables produced for selling in the two villages are similar. Most are grown in *jhum* fields and home gardens (see Box 3.7). The *jhum* cultivation system here is somewhat peculiar; normally, it is used for rice, but here vegetables are grown for commercial purpose, continuously for up to three to four years.

Changing Gender Roles in the Production of Vegetables

We observed an important change in the management of home gardens. Traditionally, men considered it 'below our dignity to tend the home gardens'. They were the domain of women. However, this mindset is changing. The community has accepted the fact that men's contribution to the maintenance of home gardens significantly increases the cash income of households. Men's efforts to enlarge home gardens and their sharing of household chores are allowing women more time to sell the produce. It has become common to see both women and men returning home with head-loads of vegetables from gardens far from home. In one discussion an elderly gentleman quipped, 'Women are weak, they cannot bring much quantity of harvest so I do the harvesting and bring twice the quantity she can carry.' Another man told us that harvesting

> **Box 3.7** **Situation for an Average Part-time Vendor in Tsiese Basa**
>
> Unlike the Pholami vendors who stay in Kohima for at least four days, the vendors from Tsiese Basa and Merema prepare a consignment sufficient for one day and return home in the evening. Costs (INR) for a typical market day are:
>
> | | |
> |---|---|
> | Transportation | |
> | From the village to Kohima by community bus | 50 |
> | To the vending site | 20 |
> | Lunch and snacks during the day | 40 |
> | Packing material | 10 |
> | One day's wage | 100 |
> | Total | 220 |
>
> An average day's sales are about INR 330, providing a (small) profit of INR 110. The vendors and village elders that we interviewed observed that, in general, selling vegetables as a business is not profitable because the quantities are very small. It could be made more profitable if production increased, and this would be possible when the fencing system is strengthened to keep cattle out of the crops. They said that regular vendors who purchase vegetables wholesale should refrain from taking advantage of inexperienced vendors. In addition, they suggested providing vendors with improved varieties of different vegetable seeds that would help increase production. Farmers are good at trying out new crops. For example, a woman vendor described their success with a variety of bean introduced by a government officer named Joshua. They named the produce 'Joshua beans'.

is best done early in the morning or late in the evening and, therefore, it is better that his spouse stays home and does the household chore while he does the harvesting.

To understand the input of men and women into home gardens, we conducted a survey of 17 households in the two villages. In all of these households, both men and women take part in all growing operations, indicating a significant change in the attitude of the men. In the past men would have contributed only to building fences and clearing the jungle. Now they also plough, sow, weed and harvest. The only activity where men remain almost absent is vending; this has remained a women's job.

Main Enabling Factors and Constraints

Compared with Pholami, these villages are much closer to Kohima, both are served by a regular bus and National Highway 61 passes through them. Vendors can collect vegetables the day before marketing them and can return home when their business is finished.

However, while transportation is not a problem, villagers face constraints in terms of production. They find it difficult to grow crops with high market demand, for example, potatoes and cabbages; they could benefit from more technical know-how. Vendors also highlighted the fact that production is affected by pests and diseases, and that management skills could be improved.

The major problem is free-roaming cattle and the very high cost of building fences, especially for the poorer households. This explains why, despite the fact that home gardens contribute a significant if not major portion to household cash incomes, they continue to be small. Their average area is about 30 m by 20 m although some are as large as 80 m by 40 m. Village resolutions and the Cattle Trespass Act passed by the Nagaland Legislative Assembly to control cattle throughout the year have little or no effect because it is the 'elite' of the village who keep cattle and they do not have home gardens. They continue to defy the resolutions and government orders. A number of extreme measures have been resorted to by village functionaries and farmers to change the situation: killing stray cattle, imposing heavy fines on the cattle owner and social fencing. Apparently, these actions have only led to conflicts between village functionaries and the cattle owners. No solutions have yet been found.

Options for Intervention

Table 3.7 summarizes the needs and options that we identified in the two villages. Most of these are interlinked. Producers and vendors are clear about the crops that have good marketing potential. They are also

Table 3.7
Options for Increasing and Improving Production of Marketable
Vegetables and Fruits in Merema and Tsiese Basa

Issue	Choices for increasing production in order of ranking	
	Merema	Tsiese Basa
Supply problem/increase in production to respond to strong demand	Papaya, orange, chilli, guava, banana	Passion fruit and leaves, papaya, mustard leaves, tree tomato
Largest quantity produced and sold	*Solanum* spp., banana, guava, orange, chilli, papaya, mustard leaves, local garlic, colocassia shoots, *chayote* leaves	Banana, ginger, papaya, passion fruit and leaves, *Solanum* spp., chilli, colocassia, *chayote* leaves, guava
'Distress sales' during peak season	*Solanum* spp., guava, mustard leaves, tomato, *chayote* fruits	*Hibiscus* spp. (*gakhro*), *Solanum* spp.
Best price per unit quantity	Orange, chilli	Papaya, guava, cucumber, banana, chilli, mustard leaves, wing beans, tree tomato, tomato, colocassia
Desire to produce in bulk for sale but lack 'know-how'	Potato, onion, cabbage, grapes, passion fruit, field pea, carrot	Potato, *raja* chilli, onion, cabbage, groundnut
Crops highly susceptible to pests and diseases	Cabbage, mustard leaves, orange	Tomato, cabbage, tree tomato, ginger, potato, groundnut

Source: NEPED fieldwork.

aware that some crops are in high demand but they lack the production know-how to respond properly and in a timely manner. To enable farmers to increase the production of high-potential vegetables while maintaining their organic production methods, we initiated a dialogue between research laboratories and producers, especially in the context of integrated pest management, bio-fertilizers and bio-compost. We encouraged the sharing of good or innovative practices among farmers based on concrete examples of other farmers in the region, such as the case of potato cultivation in Viswema village.

Khonoma

Khonoma is 22 km west of Kohima and is inhabited by people of the Angami tribe. There are 391 households with a population of about 2,180 (Cairns 2004). The village is easily accessible and connected to Kohima by a daily community bus. Vegetable vending in Khonoma began with the introduction of the first bus route to Kohima in 1981. Before that, a few women sold wild vegetables, such as banana leaves, when they travelled to town on foot for social visits. The community buses have greatly enhanced mobility. Villagers go to town not only for social visits but also for economic reasons, marketing agricultural or forest produce and buying essential items for the family.

There are 25 vendors who produce vegetables in their own fields and sell them at the market. Sometimes they sell their own produce; when prices are good, they sell to regular retailers. On a typical marketing day a vendor spends about INR 200 to cover expenses and usually makes a small profit. Most vendors said that they are willing to sell their produce in the village at wholesale rates if the prices are reasonable. All women vendors interviewed feel that vegetable selling is still a profitable business. It allows them to supplement the family income to pay for children's education and to buy clothing and other essential household items such as milk, sugar, meat and even rice if household production is insufficient. Most said that bulk production for commercial sale could lead to marketing problems unless better linkages can be established. They also noted that a labour shortage could arise because all the children go to school and do not help in the field (except during vacations).

Potential Interventions

We heard several ideas from vendors for improving their situation. One part-time vendor suggested that the KTC should enforce strict use of designated marketplaces. She said, 'Today, vendor produce is seized, the next day nothing happens. So vendors are encouraged to occupy the footpath again with a bit of risk and insecurity. The public should be disciplined to go to the designated marketplaces and not buy from the footpath.' Another commented, 'The government should provide good quality seed and impart technical know-how of producing off-season vegetables that have a good price.'

We suggested that an appropriate marketplace be provided to vendors for a nominal fee. Vendors think that the old Nagaland State Transport (NST) complex, currently occupied by cobblers, would be a good location. However, when we communicated the vendors' idea to the administration, they disagreed and argued that such a centrally located place for vegetable marketing is unacceptable from a hygienic and sanitation point of view. Further research is needed on this option.

Another suggestion we put forward was to establish a collection centre or storage facility in the village from which the produce can be taken to wholesalers in Kohima by self-help groups or cooperatives. Farmers could expand the area under vegetable cultivation and produce in bulk for commercial purposes.

Another option is to organize more experience-sharing tours among progressive farmers to exchange more information about good practices. Successful farmers could be duly acknowledged and given small incentives—in cash or kind—by government development departments.

Asked if men could sell vegetables to overcome the harassment being faced by women when they are loading and unloading produce, one of the vendors responded:

> It is a role not expected of men to do because traditionally men and women have specific assigned roles. Men would instead do cane and bamboo work, stone work or carpentry for household use or earn additional income to the supplement family income. They may not have the patience to sit on the roadside the whole day selling 'pity' [seen as of low value] vegetables or may not know how to sell. They may not earn anything at all. Even if they manage to sell all vegetables, men may not know what items to buy back home. Marketing cash crops in bulk is, however, done by men as it is to be sold at distance market places like Dimapur. Men are more mobile and can be away for a longer period of time unlike women. Loading and unloading bulk produce is also easier for men to handle than women.

CONCLUSIONS

Our study of the diverse vegetable production and marketing practices of women vendors points to a number of enabling factors. Soils, rainfall, geographic location and elevation allow Nagaland farmers to grow numerous crops simultaneously in the same location with comparative ease. They can produce crops under natural conditions without irrigation or chemical fertilizers. The vegetables they sell in Kohima are of two sorts: wild and cultivated. Wild vegetable are mostly collected from the primary forest, whereas cultivated ones have three sources: *jhum* fields, TRC fields and home gardens. The many varieties offered suit the tastes of various customers well. Most women vendors have good support from their family and community. While they are away (selling), men take care of household chores. The community provides support through the system of social fencing. Although vendors and vending are formally regulated by KTC staff by means of regulations and standing orders (for example, evacuations), in general they receive sympathy from these important decision makers.

However, marketing is not without hardships and constraints. The common constraints are limited production to attract wholesale dealers from towns and cities, and insufficient quantities of vegetables brought to Kohima to venture into wholesale trading by part-time vendors. Merema and Tsiese Basa face the problem of roaming cattle and the inability of producers to pay for fencing. Vendors from Pholami village are hampered by poor transportation facilities. More (research) efforts should go into finding alternative marketing chains to deal with these multiple problems (see Box 3.8).

The issue of marketing sites continues to plague all stakeholders. The KTC has prepared the infrastructure for vegetable vendors, but the vendors do not use it. They prefer to squat at any convenient place and carry on their business freely, causing inconvenience to pedestrians, traffic and shopkeepers. The vendors argue that the infrastructure provided is useless because no customers come to buy vegetables in this area. This seems to be a 'chicken and egg' problem.

To deal with the constraints, we are exploring a number of intervention strategies adapted to the local context. In the case of Merema and Tsiese Basa we have encouraged more rigorous enforcement of the Cattle Trespass Act and village resolutions prohibiting cattle from 'free-grazing' during winter. In Pholami we increased motivation and awareness of the benefits of producing and selling vegetables. We also encouraged a community initiative to secure a truck to travel between the village

> **Box 3.8** **An Alternative Marketing Chain: The Vendors of Pfutsero**
>
> Farmers told us about a group of women who are trying an alternative marketing method and we went along to investigate.
>
> Growers bring their produce by head-load from neighbouring villages to these women at a collection centre in Pfutsero. The group deals only in select vegetables, mostly from the garlic family. The women package the produce according to the requirements of their regular customers. They hire a truck and proceed to Dimapur.
>
> On the way to Dimapur, their regular customers wait for the arrival of the goods at the roadside in Piphema and Medziphema. The women do not even get off the truck but toss the packages of vegetables to the customers shouting, 'That will cost you INR 700. I will come tomorrow and collect the money.' Then they continue their journey.
>
> On arrival in Dimapur this day, there are no waiting customers. When asked what they were going to do about the vegetables, the women replied, 'Today we arrived a little too late to do business, so we shall store them in our godown [which they have rented]. By tomorrow morning when the community buses arrive, all will be taken.' The following morning all the produce left in storage is sold.

and the nearest town. We provided or suggested technical and material support for farmers to cultivate vegetables that are in demand, but for which they lack knowledge; for example, the farmers of Tsiese Basa and Merema, who want to cultivate potatoes organically but lack the management expertise. Integrated pest management training in this area would augment production.

The Mao residents of Kohima initiated the Mao Market Complex. They observed the Mao women vegetable vendors coming from villages with their vegetables and having trouble eking out a livelihood. They discussed their observations in their own community in Kohima, which then purchased a building in a centrally located place. This market complex is vibrant with women vendors from the Mao community and has solved the problem of location for the women of the tribe. Learning from this success, we created an informal forum of town decision makers to draw up a plan for a site for vegetable vending. A start has been made in this direction, but more time is needed to produce a fruitful outcome.

REFERENCES

Cairns, M. (2004). Fuel and Fertility: Alder's Role in Land Use Intensification. Presented at Panel Discussion on Shifting Cultivation (*Jhum*): Policy Imperatives for North East India, 7 July 2004. (Reproduced by LEAD-India in Partnership with ICIMOD.)

Government of Nagaland (2001). *2001 Statistical Handbook of Nagaland.* Kohima: Directorate of Economics and Statistics.

Supong, K. (1999). Farmers' Knowledge of Shifting Cultivation in Nagaland. Report submitted to IDRC, Agriculture Research Station, Mokokchung, Nagaland.

4

Enhancing Farmers' Marketing Capacity and Strengthening the Local Seed System

Action research for the conservation and use of agrobiodiversity in Bara District, Nepal

Photo credit: Deepa Singh

DEEPA SINGH, ANIL SUBEDI AND PITAMBER SHRESTHA

Background and Context

The Country and the Research Site

Nepal is a mountainous country; it extends from the Indo-Gangetic plains in the south (60 m above sea level) to the high peaks of Mount Everest (8,848 m above sea level) in the north. The country is divided into five ecological regions: the Himalayas (over 4,000 m above sea level); high mountains (2,000–4,000 m); mid-mountains (1,500–2,000 m); the Siwalik range (300–1,500 m); and the extension of the Indo-Gangetic plains in the western part of the country, also know as *terai* (less than 300 m). The climatic conditions range from tropical in the south to freezing alpine in the north.

Nearly 81 per cent of the people of Nepal rely on agriculture for their livelihood (CBS 1999). Farming is subsistence oriented; landholdings are small and fragmented, usually less than 1 ha per household, and families are large (average of 5.6 people per household). Productivity is generally low at less than 2 t rice/ha. Farming is labour intensive and women play a major role in farm activities. On average, they contribute up to 75 per cent of total agriculture labour (WFDD 2002).

Kachorwa, where the research described in this chapter was carried out, is a small village in the Bara district (Figure 4.1). It is situated in the Indo-Gangetic plains and experiences a subtropical climate. The mean annual temperature in the area is 24°C; January is the coldest month and May the hottest (35.4°C). Kachorwa is typical of the settlement pattern in the *terai*, where clusters of houses are surrounded by agricultural lands. The settlement at Kachorwa is an old one and can be reached by a rough pebble road. In the village the road is only usable during fair weather. The 914 households in the village comprise five ethnic groups: Brahmin-Chettri at the top of the social hierarchy, Sah, Koiri, Muslim and the lower caste or 'untouchables' at the bottom. The lower caste, which represents the most disadvantaged group, is socioeconomically and politically marginalized.

Both nuclear and joint families exist in Kachorwa, although there are slightly more nuclear families. The culture is typical Indo-Aryan, in which males are dominant and females are encumbered by taboos. Most women are confined to the house, especially in the richer farm families.

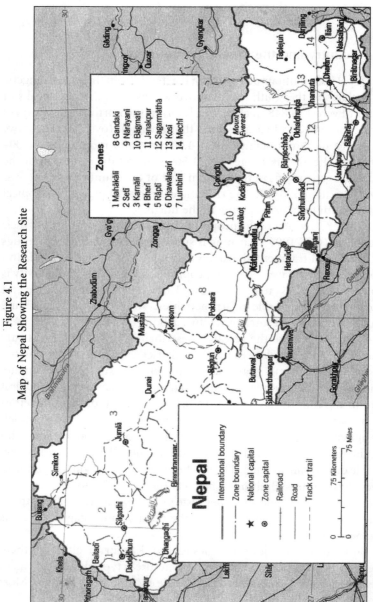

Figure 4.1
Map of Nepal Showing the Research Site

Source: Perry-Castañeda Library Map Collection.

The people of Kachorwa depend on a combination of on- and off-farm activities for income. Most engage in subsistence farming, usually growing three crops a year.

The level of interaction between villagers and the research and extension system is high; there is also contact with the wider Indian region across the border in terms of technologies and agriculture inputs. This is in contrast to many other parts of the country and makes this case study particular. Cultivated land is mostly *khet* (irrigated and rainfed) land; *bari* (non-irrigated) land (*bhitha*) is found in only a few villages. Kachorwa is rich in terms of crop diversity: 18 crops and 34 vegetables are grown. Species diversity is also very high; in the case of rice, 24 different varieties are grown.

Link to the 'In-Situ Crop Conservation' Project

Bara is one of three sites in the 'In-Situ Crop Conservation' project—a global project implemented in nine countries and coordinated by the International Plant Genetic Resources Institute (IPGRI). The Nepal Agricultural Research Council (NARC) and Local Initiatives for Biodiversity, Research and Development (LI-BIRD) have implemented the Nepal component of the project in partnership with farming communities and groups in Bara, Kaski and Jumla districts, which represent plains, mid-hills and high hills areas respectively.

The first phase of the project (September 1997 to December 2001) was funded by the Netherlands Development Agency (NEDA). Three main objectives were: strengthening the scientific basis of *in situ* conservation on-farm; capacity building of national institutions in conducting on-farm conservation; and adding value to local plant genetic resources for the direct benefit of farmers. Considerable effort was made to include social and gender issues while planning and implementing project activities in the field. Several specific results have been achieved in the broad areas of assessment and adaptive management of local crop diversity on-farm; capacity building and creating representative partnerships; and mainstreaming the use of local crop diversity in agricultural development. Space constraints prevent us from providing more details (but see reports cited in the reference list—Gauchan 2000; Rana et al. 2000; Subedi et al. 1999; Subedi et al. 2001).

At the end of the project period many activities in and with farming communities were still in progress. Gaps and new needs were also identified. IDRC agreed to fund a second phase of the project for two years beginning in September 2002. Our research on social and gender analysis (SAGA) complements the second phase, particularly by providing a better understanding of the social and gender dimensions of rural livelihood and by strengthening grassroots organizations interested in improving the local seed system. Our research is action oriented and has three broad objectives:

1. enhancing the capacity of farmers at different socioeconomic strata and gender groups in seed production and marketing, thereby contributing to their economic improvement;
2. strengthening local seed marketing systems and networks; and
3. enabling government and non-governmental organizations (NGOs) to mainstream social and gender perspectives in agriculture research and development.

This represents a comprehensive and challenging agenda and, in this chapter, we highlight some of the work in progress.

Key Natural Resource Management Issues

Farmers in Nepal have been maintaining biodiversity to meet the diverse needs of their households. Many landraces of the region have unique characteristics; for example, Bhathi is an excellent rice that performs better than modern varieties under waterlogged conditions. In many places, landraces are the only option for the marginal lands the farmers rely on and they have contributed substantially to food security, especially of the resource-poor households in Kachorwa (Rana et al. 2000). Many rice landraces have been identified to have economic and social value, and they are in great demand. At the same time, lack of market information about their high-quality attributes has limited the supply of seed and grains of such crops, for example, Lalka Basmati and Kariya Kamod rice. Farmers also lack certain skills and knowledge about production techniques to use with quality seeds of local landraces (Gauchan 2000). Generally, farmers save their own seed from year to year, but research has shown that this method of seed selection and conservation could be improved. This is one of the tasks we identified as part of our work.

In Nepal landraces are exchanged through informal seed networks. However, the poor links between small-scale seed producers and market channels are considered a major constraint. Farmers have not been able to maximize the benefits of the biodiversity they maintain in their fields. Because modern varieties have higher average yields and fetch a better price, farmers are inclined to plant introduced varieties. Thus, there is a need to build the capacity of the farming communities to produce quality seed and to enhance their skills in marketing to strengthen the local seed supply system and seed marketing networks. A second step is to respond adequately to consumer demands and to investigate the prices of landrace seeds.

Farmers do not conserve biodiversity for the sake of conservation. However, conserving biodiversity is important for the whole society, and it is thus fair that farmers benefit from their efforts. Providing market options can be among the least expensive conservation instruments (Brush 1991). Helping market local crops and linking small-scale seed producers with markets could be effective ways to benefit farmers.

Integrating Social and Gender Analysis into Natural Resource Management

SAGA helps us understand the roles that different members of household groups and families play in natural resource management. The management of agricultural biodiversity involves a series of decisions about planting, managing, harvesting, processing, exchanging and acquiring seeds that affect the process of evolution and adaptation of crops to their environment (Jarvis 2000). Understanding how farmers make decisions about the management of biodiversity has gender dimensions; the work and responsibilities men and women undertake differ, resulting in diverse experiences and distinct knowledge (Adamo and Horvoka 1998; Subedi et al. 1999).

The seed system in Nepal encompasses a wide range of activities, and a particular person performs each activity. The division of roles and responsibilities has been built up according to the knowledge and skills of women and men, which have been passed on from one generation to the next. Seed production and the management of biodiversity are truly *social* phenomena.

Access to resources and knowledge vary between groups of different social status, giving rise to inequities in social and gender dimensions. To a large extent, the resource-poor households rely on resource-endowed farmers for genetic materials through farmer-to-farmer networks, which often extend beyond the village boundary (Subedi et al. 2001). Farmer-to-farmer networks play a significant role in the dissemination of genetic materials (seed and planting materials) as well as associated knowledge. In this type of system a (usually small) number of people of higher social status introduce and distribute seeds. We have termed them 'nodal farmers' (ibid.). Understanding the social relations in the networks is an important first step toward strengthening and expanding their function; for example, to increase the efficiency and effectiveness of the dissemination of technology. This, in turn, can contribute to the early realization of benefits accruing from adoption of the technology by different categories of farmers.

Our research aimed to address the following broad questions: What is the existing situation in terms of consideration of social and gender perspectives in agriculture research and development? What are the constraints to integrating social and gender concerns into agricultural research and development? In addition, specific research questions were more related to the seed system in Bara: What are the roles and responsibilities of different social and gender groups in seed production and marketing? What is the existing seed flow mechanism? How can the existing seed marketing systems and networks be strengthened?

Fieldwork

Review of Literature

Two literature reviews were carried out in the initial stages of the study: one on existing seed production and marketing methods (for example, the Nepalese seed regulation policy of the Ministry of Agriculture and proceedings of workshops hosted by the Nepalese Seed Board), and the other on mainstreaming social and gender concerns in agriculture research and development (for example, NARC policy documents, Ministry of Agriculture policy and planning documents, the Nepalese government's Five-Year Plans). We paid particular attention to how SAGA has been mainstreamed by NARC in research and by the Ministry of Agriculture and Cooperatives (MOAC) and the Department of Agriculture (DOA) in policy and extension. We tried to identify limiting factors and opportunities to facilitate the effective integration of social and gender perspectives into agriculture research and development.

Interactions with Government and Semi-government Personnel

Based on the literature review, we identified agriculture research and development organizations and institutions that have taken on the responsibility of mainstreaming social and gender concerns. A series of meetings, personal interviews and focus group discussions were conducted with personnel from the outreach division of NARC and the Women for Development Division (WFDD) of MOAC. Gender focal point people and members of gender working groups of MOAC were also contacted.

A focus group discussion was conducted to document past efforts toward integrating social and gender perspectives into agriculture research and development. An in-house discussion among LI-BIRD project staff was held and a checklist of questions was developed regarding the current status of these efforts as well as constraints and enabling factors.

Our interactions with government staff were problematic. Getting appointments was difficult. We also found that most people we interviewed had no clear understanding about gender or that other, sometimes contradicting, ideas exist. Conversations repeatedly tended to focus on women's issues rather than gender issues, forcing us to bring the topic back on track, which was not always appreciated by the higher authorities whom we interviewed. Our sample group was limited and, therefore, we cannot generalize about this situation. However, the experience made us more aware of the challenge, the on-the-ground realities, and the processes playing out in policy-making spheres. It also intensified our own attempts to pay attention to and integrate social and gender issues into our field research.

Rapid Assessment of the Market

Various participatory research tools were used to acquire insight into the existing rice seed marketing systems. Questionnaires were developed and a series of market studies was conducted. Focus group discussions were carried out with nodal and other farmers, followed by key informant interviews to document existing knowledge and the skills of both women and men. A detailed market survey was conducted with a total of 98 respondents (representing different ethnic groups, socioeconomic strata and sexes). Attendance at the local weekly markets (*haat* bazaar) allowed us to observe directly the trends in seed transactions. These methods proved useful and generated valuable insights into the everyday realities of farmers and vendors, their interests, aspirations, problems and constraints.

Training in Mass Seed Selection

A training manual and posters on rice seed selection were prepared. The training was conducted in two phases. The first included an initial orientation before planting and training in seed selection in the standing crop. Quality control, packaging and labelling were all covered during the second phase. Resource people from the National Rice Research Station and Regional Seed Laboratory were invited to give the training, which was highly appreciated by the participating villagers.

Stakeholder Workshops

Workshops were held for various stakeholders, such as seed producers, agro-vets, rice seed sellers and grain sellers within and outside the village to disseminate information about ongoing activities and to assess the demand for local rice seed.

During fieldwork, especially in Kachorwa, we experienced several constraints. As mentioned, Kachorwa is a typical *terai* village. Villagers speak the Bhojpuri dialect and Hindi is understood to some extent;

thus, language was one of the barriers, although interactions were manageable. The cultural barrier proved to be much more of a challenge. Interviewing women was very difficult, as they are not allowed to interact with 'outsiders'; thus, repeated visits were necessary to build rapport and make the women feel comfortable speaking out. In addition, men were uneasy speaking to women from 'outside'. Interviewees were sensitive about answering questions about decision making. We dealt with this by explaining gender issues repeatedly and asking them if they understood our questions.

SOCIAL AND GENDER PERSPECTIVES IN AGRICULTURAL RESEARCH, EXTENSION AND POLICIES

Getting Social and Gender Issues into the Mainstream of Agriculture Research

NARC is a leading, autonomous centre for agriculture research in Nepal. Its major challenge is to develop technologies that will enhance or support high and sustainable economic growth, alleviate poverty and ensure food security in the country. This goal can only be achieved by developing technologies that match the socioeconomic needs of the farmers and are socially acceptable and viable. To address this challenge, socioeconomic research was initiated in 1977 by the agronomy division. However, social and gender studies are currently conducted by the outreach division, based in Khumaltar, and in some regional and area-specific agricultural research stations within the country.

Social research is not yet strong in Nepal. Sound social analysis prior to the development of any technology has yet to be put into practice. We learned this during interviews with government staff by discussing a number of examples. For example, a new zero-tillage system and a rice-planting machine that were developed and promoted by NARC have both positive and negative impacts. They reduce the overall workload of farmers substantially, but at the same time reduce employment opportunities for those who work as labourers.

Another example is the hybrid varieties that are being introduced in all major crops. These are expensive and require an optimum environment and inputs. Hybrids may be high yielding, but may not be suitable for all categories of farmers or for the niche agro-ecology of the country. We initiated a joint research effort between NARC and LI-BIRD to address this shortcoming by identifying a wheat variety that is suitable for rain-fed conditions. Such a variety would require less water and be more suitable for poor farmers who cannot afford irrigation facilities. Another socially conscious initiative developed by the national research system is a set of mushroom cultivation practices that targets different categories of farmers instead of a single model package for all. What is important in both these examples is that actual research work takes place at the farm level with farmers involved. This approach is not as easy as it sounds, and we still have a lot to learn about how to do it well, but it is central to LI-BIRD's philosophy.

NARC's new 20-year strategic plan states that the centre's future research will take gender concerns into account. The plan mentions a rising concern about clients and their needs, and the need to increase the rate of adoption of new technologies. In past years some studies focusing on gender aspects have been carried out by the outreach division, but they were confined to the central level. The outreach division identifies needs and formulates programme activities at the so-called 'outreach sites', and some women participate in these outreach activities. Some simple technologies were developed to meet women's requirements, for example, for fruit thinning and processing. Women are also involved in such activities as visits and on-farm trials conducted by NARC.

NARC has also been including gender concerns or women's issues in the development of research proposals, and it has carried out studies of the impact of new technology on male and female beneficiaries. For example, when milk marketing opportunities were created, men took over the sale of milk from women, and the resulting cash income was spent mainly on men's personal needs. However, NARC scientists perceive this area of research as a separate programme activity rather than a core part of its work.

NARC and LI-BIRD are working together on another initiative in participatory plant breeding. Special attention is being paid to women in participatory technology development, particularly in the development of a new maize variety. In Gulmi, with LI-BIRD's support, Resunga

Composite, a new lodging-tolerant variety of maize, has been developed by a woman farmer-breeder. This very successful variety is now in the process of being released by the national system and represents a good example of strong and meaningful farmer participation. Through NARC–LI-BIRD collaborative efforts, we hope to demonstrate how NARC scientists can change their conventional practices. Working with women and men farmers on their farms is a crucial element in our way of doing research.

Social and Gender Perspectives in Extension Work

Nepal's agriculture extension system has made several attempts to incorporate gender concerns. However, other social factors or perspectives are still ignored. The implementation of extension programmes is mainly based on simplistic socioeconomic categories, such as rich, intermediate and poor. Different programmes are designed and implemented for each category of farmers in a community, but in practice there are many problems. No specific or detailed needs assessments are made, and, most critically, the very poor farmers have been and continue to be ignored by the system.

In Nepal all so-called gender-focused activities are centred on women. Instead of integrating gender perspectives into the system, separate activities are implemented for women. The inclusion of a certain percentage of women in various agricultural development activities was made compulsory in the extension system's five-year operational plans, beginning at 25 per cent in the Eighth Five-Year Plan (1992–97) and increasing to 35 per cent in the Ninth Plan (1997–2002). Currently, it is set at 40 per cent in the Tenth Five Year Plan (2002–07).

There are numerous special programmes for women. In addition, at the policy level the inclusion of the maximum number of women in pro-women technology development is promoted: for example, in vegetable production, food preservation and integrated pest management (IPM). Income-generating activities for women farmers—in vegetable production, bee-keeping, agro-processing and marketing—are carried out every year by district offices. District agricultural development offices (DADOs) conduct separate training sessions exclusively for women farmers; they also include women in regular activities.

DADOs carry out gender matrix analysis to select participants for farmers' field school in post-harvest, irrigation and IPM practices. Women IPM trainers have been identified using this gender matrix analysis tool. In Bhaktapur women are conducting trials on the management of potato tubers. Rewarding women farmers based on their performance and contributions in agriculture is another DADO practice.

The MOAC also recognizes women's contribution to agriculture. For example, on World Food Day, the WFDD has been recognizing the achievements of women farmers for their work in agro-processing in five development regions. Women farmers are also encouraged to become involved in marketing. The WFDD has been providing them with market stalls in the Kalimati vegetable market. In Kathmandu some women farmers are providing fertilizer to other farmers. The WFDD includes women farmers in problem census and problem surveys and in its annual district-level programme planning. However, this is not widespread.

In summary, a start has been made toward the inclusion of social and gender perspectives in agricultural research, extension and development. However, several serious constraints still exist. Factors at the policy, implementation and grassroots levels narrow the possibilities for addressing social and gender issues in a meaningful way.

Changes in attitude play a major role in the entrenchment of SAGA. Ignorance about the importance of social and gender aspects in planning, development and dissemination has resulted in the inefficient implementation of most policies. Existing social norms and culture influence the involvement of women and men in research and development. For example, cultural norms in the *terai* are very conservative; women here are confined to the household area. Technology dissemination and action research are hindered by this.

The personnel involved in research and extension at all levels are profoundly technical and many ignore the social aspects of developing new technology. Their lack of exposure to concepts, tools and methods for SAGA is a major setback for effective mainstreaming of social and gender concerns in agriculture research and development. The lack of conceptual clarity on social and gender issues and the failure to understand the importance of SAGA in programme identification, design and implementation are blocking effective mainstreaming.

To provide an example of how SAGA can be solidly incorporated in development research, we now turn to our case study of Kachorwa, Bara.

SOCIAL AND GENDER ANALYSIS OF THE KACHORWA SEED SYSTEM

Importance of both Modern and Local Varieties

Farmers of Kachorwa grow both landraces and modern varieties of rice to meet their diverse needs. Landraces have sociocultural value, are adaptable to niche environments and are favoured for their taste. Landraces are grown for different purposes by farmers of different socioeconomic categories, depending on their needs. Some landraces such as Sathi (Gamadi rice) are grown by all farmers—rich, intermediate and poor—in small parcels of land. This is an essential variety for the *Chath* festival, an important and widely celebrated festival in the *terai*. Other landraces such as Lajhi, Basmati and Kariya Kamod are grown for making other products like *khir*. Aromatic rice varieties, such as Lalka Basmati and Kariya Kamod, are grown for their good quality. These latter varieties are grown by the rich and intermediate farmers and are used exclusively for entertaining guests and for specific festivals. Some landraces such as Bhathi and Nakhi Saro are grown for their adaptability.

The rich farmers own larger parcels of land and have more options in choosing a rice variety to grow. Modern varieties are preferred over local varieties by all categories of farmers because of their higher productivity; taste and marketability are also important factors. Wealthier farmers appreciate the non-lodging character of certain modern varieties. Rich and intermediate farmers hire labour to work on their farms and pay labourers with varieties like Jaya and Sabitri. Farmers' choice of modern varieties also depends on such other parameters as straw quality, agro-ecological adaptation, cooking quality and milling characteristics (percentage undamaged grains). We found that these latter criteria are more important to women farmers and they relate closely to women's activities. Landraces are chosen by farmers in all three socioeconomic groups because of the role they play in religious functions. Rich and intermediate farmers also appreciate the importance of certain landraces for social functions, such as entertaining and festivals (Figures 4.2 and 4.3).

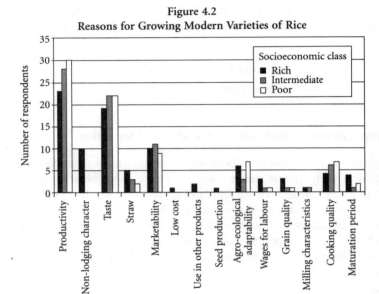

Figure 4.2
Reasons for Growing Modern Varieties of Rice

Figure 4.3
Reasons for Growing Landraces of Rice

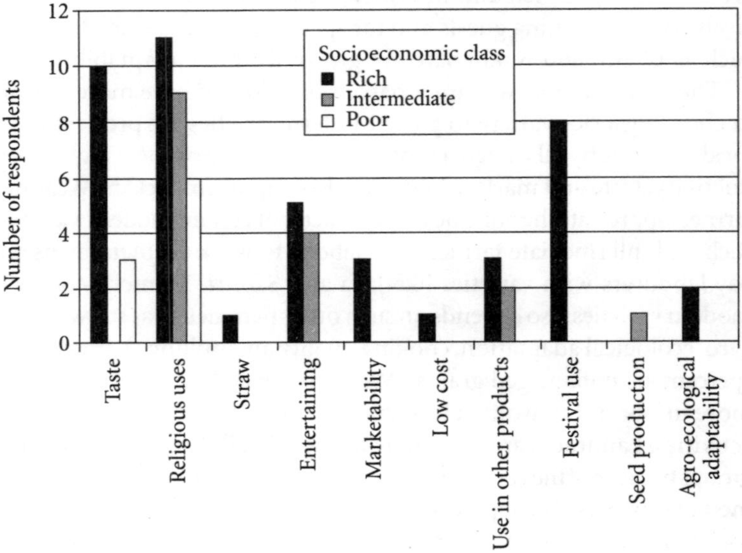

The importance of seed purity of both local and modern varieties was well understood by all farmers. Better production was the main reason for maintaining seed purity; other reasons included uniform maturity, less pest infestation, consumer preference and better tillering.

Seed mixtures are a major problem in the case of landraces and they decrease the quality of the crop. The introduction of modern varieties necessitates the purchase of new seed for each crop or year. Mixing of seed due to out-crossing and mechanical mixing is the most prevalent problem perceived by most farmers. However, the rate of seed replacement and the reasons for seed replacement vary with social class and seed variety. Off-types are a more serious problem in the aromatic rices, such as Lalka Basmati, and these seeds are replaced every year. In the study area the rate of seed replacement varied from every year to once in six years. Seed replacement is more frequent among richer farmers, whereas an interval of three years was most common for intermediate (31.3 per cent) and poor (32.4 per cent) farmers (Figure 4.4).

Figure 4.4
Period of Seed Replacement for Three Socioeconomic Categories of Farmers

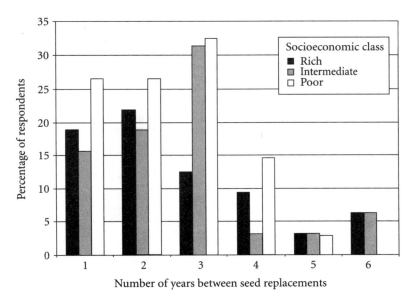

Seed Sources

In Kachorwa richer farmers have better access to seeds within and outside the village, which allows them to buy new seeds more frequently than intermediate and poor farmers, who do not have this kind of access.

Our fieldwork revealed that the informal seed system is predominant in terms of obtaining both landrace and modern seed varieties. However, most farmers depend more on the informal system for landraces, whereas there are other options for modern varieties. Most farmers in all three socioeconomic categories save seeds of both landrace and modern varieties for planting the next year. Exchanging seeds with neighbours and relatives, and purchasing seeds from agro-vets were other methods for acquiring landraces. Rich and intermediate farmers commonly bought seeds from agro-vets in and outside the village as well as from the Indian markets. Intermediate and poor farmers depend more on their own saved seeds and on seeds obtained through exchange locally. They find seeds sold by agro-vets expensive and in some instances they are not available on time. Poorer farmers also obtain seeds in exchange for their labour (Table 4.1).

In terms of seed quality, most farmers perceived that self-saved seeds are more likely to be pure than those obtained from relatives and neighbours. Other reliable sources they mentioned frequently were

Table 4.1
Sources of Seeds across Wealth Categories

Source	Landraces			Modern varieties		
	Rich	Intermediate	Poor	Rich	Intermediate	Poor
Save from own crop	15	19	5	63	65	44
Relatives	3	1	1	7	8	4
Neighbours	5	2	4	8	1	4
Exchange for labour	–	–	–	4	2	4
Save + neighbours	1	–	–	1	–	–
Save + relatives	–	–	–	–	1	–
Agro-vets	–	–	–	10	8	5
NRRP	–	–	–	5	2	1
Indian markets	–	–	–	10	7	3

Note: NRRP = National Rice Research Programme.

agrovets and government farms; however, these sources deal only with the modern varieties. NARC's National Rice Research Programme is another source for modern varieties, and Kachorwa is situated close to a NARC research station.

The resource-rich farmers play an important role in seeking material from different sources, both within and outside the village. Nodal farmers play a significant role in trying out new varieties and facilitating subsequent seed flow (Subedi et al. 2001).

Another source of seeds is the weekly market, called the *haat* bazaar. Farmers bring small amounts of seed, especially landraces, to these markets to sell just before the planting season, although this practice is more common for vegetables than for rice. Seeds are also obtained from temporary shops (known as *beez bhandar*) before and during planting season. These are usually operated by richer farmers, who can afford to bring seeds and fertilizers from India. Figure 4.5 summarizes our findings graphically.

Figure 4.5
Existing Seed Marketing Channels for Kachorwa Farmers

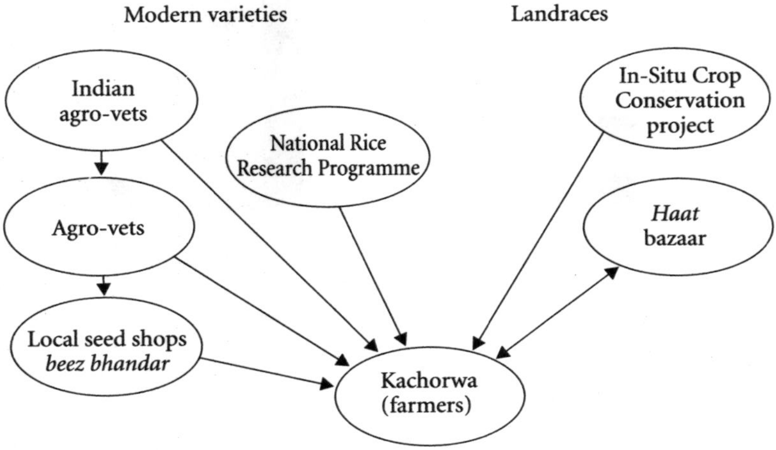

Modern varieties   Landraces

- Indian agro-vets
- National Rice Research Programme
- In-Situ Crop Conservation project
- Agro-vets
- Haat bazaar
- Local seed shops *beez bhandar*
- Kachorwa (farmers)

Source: Focus group discussion, 2003.

Types of Exchanges

The farmers at the study site engage in two types of exchange: seed-for-seed and seed-for-grain. For immediate transactions the ratio is 1:1 for both types of exchanges. However, the ratio is 1.5:1 when payment is delayed, for example, until after the crop is harvested. Poor farmers often must accept this higher ratio, because they do not have surplus seed to exchange and must wait to harvest the current crop. In other words, they pay more for their seed. Rich farmers usually have sufficient stored grain to take advantage of the lower rate (Figure 4.6).

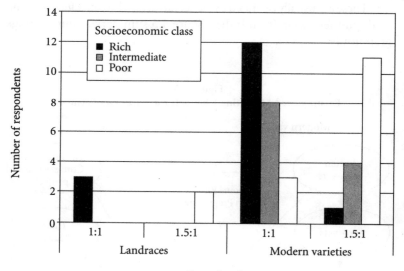

Figure 4.6
Rates for Seed Exchange across Socioeconomic Categories

Seed Selection Methods

Farmers adopt various seed selection methods to maintain the purity of their crops. The methods are the same for both local and modern varieties. The selection process can be divided into five categories that take place during harvest and storage (Figure 4.7).

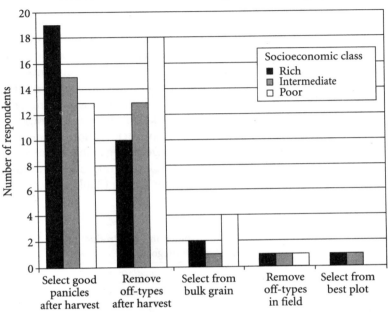

Figure 4.7
Seed Selection Methods According to Wealth Category

The removal of off-types from bundles of harvested rice is the usual method of poor farmers. Removing off-types in the field and selecting seed from bulk grain are also common.

However, the seed selection process also depends on the amount of seed required and the cultural value of the variety. Critical and careful seed selection is carried out for varieties used for religious purposes. If only a small amount of seed is needed, farmers can select good panicles from the field. When a large quantity is required, farmers will generally simply remove off-types from bundles of grain as this method is less labour intensive and time consuming.

Seed selection practices vary with socioeconomic status. Poor farmers opt for the easier method of removing off-types from the bundles and saving the remaining seed. Richer farmers are able to hire labour for panicle selection. Interestingly, poor farmers said that they often do not spend much time selecting seed from their own crops because better quality seeds can usually be obtained from richer farmers.

Gender Roles in Seed Production and Marketing

The roles of women and men in decision making were explored for the various socioeconomic categories. Among rich farmers the selection of the seed variety was male dominated, whereas it was more a mutual decision among intermediate and poor farmers. Men play a dominant role in panicle selection among rich farmers; women are also involved in the intermediate and lower categories. Both men and women carry out panicle selection, depending on the availability of men during the specific season. Post-harvest operations like threshing and storage are exclusively female oriented, irrespective of socioeconomic category. Recently, the seasonal exodus from the village of men in the intermediate and poor categories has changed gender role dynamics and is resulting in greater women's involvement in seed selection (Table 4.2).

Table 4.2
Involvement of Men and Women in Seed Production across Wealth Categories

Activity	Rich			Intermediate			Poor		
	Men	Women	Both	Men	Women	Both	Men	Women	Both
Variety selection	18	2	12	13	0	20	11	4	17
Intercultural operations	24	1	7	21	0	12	12	3	17
Harvesting	25	1	6	18	1	14	5	6	21
Panicle selection	21	0	6	19	1	11	6	4	22
Selection after harvest	26	0	4	17	0	16	10	4	18
Post-harvest operations	12	14	4	5	11	16	2	25	5
Storage	3	12	5	0	23	6	2	24	2

Use of Women's Labour

More hired labour is used by farmers in the rich category than by intermediate and poor farmers. The conservative norms and culture of Kachorwa dicatate that women are not permitted to work outside the house, and these norms are adhered to by those in the higher socioeconomic category. However, rich farmers can afford to hire labour and the usual practice is to hire the poor farmers (Figure 4.8). Looking at the poorest households, women and men contribute more or less equally (at least in numbers) to harvesting and panicle selection, while women are mostly responsible for post-harvest work (Table 4.2).

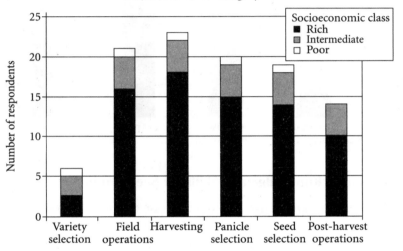

Figure 4.8
Labour Used at Various Stages of Seed Production by Socioeconomic Category

Marketing

Only rich farmers are involved in seed transactions. Within this class, the decision about the amount of seed to be sown is made by men and to some extent by women, whereas decisions about where to sell the seed and the price are more influenced by men. Men and women are both involved in the actual selling: men are involved in seed transactions both in and outside the village; women are only involved when the seeds are sold from their own home (Figure 4.9).

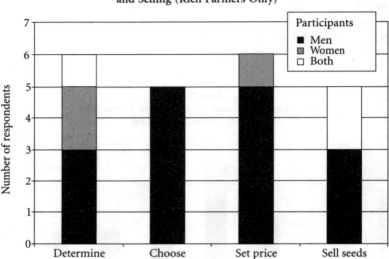

Figure 4.9
Participation by Men and Women in Decision Making
and Selling (Rich Farmers Only)

Acquiring Knowledge and Skills for Seed Production

We found that knowledge of seed production techniques and skills was fairly consistent across wealth categories, ethnic groups and sexes. Knowledge and skills are passed from generation to generation, mostly by demonstration. Knowledge was acquired mainly from household members, but farmers also learned by observing neighbours and relatives.

Capacity Building and Strengthening the Existing Seed Market System for Local Landraces

Our findings suggested that the usual selection process is insufficient for the production of high-quality seed, especially for landraces like Lalka Basmati and Kariya Kamod. Off-types are a severe problem in the case of aromatic rice landraces due to admixture of the seed and out-crossing. The morphology of aromatic rice landraces makes them more likely to out-cross; thus, selection processes must be continuous to maintain good quality seeds.

Increasing capacity in the community for seed selection and marketing was our main action strategy. To this end, we developed a training manual that explains the scientific basis of seed selection in simple language with sketches. Both men and women farmers of different socio-economic categories and ethnic groups were trained in seed selection.

The training was conducted in two stages: orientation and field-level seed selection. The first phase focused on the importance of seed selection and basic requirements, and was conducted before the rice season. The farmers were made familiar with various steps in seed production. At the end of this phase we identified farmers who were interested in putting into practice what they had learned immediately.

We also assessed the demand for particular landraces and found that farmers showed strong interest in three. Collection of source seeds of these three landraces, quality testing and seed distribution were carried out by the Agriculture Development and Conservation Society (ADCS), a community-based organization of women and men farmers of the study area, with the technical assistance of LI-BIRD staff.

Follow-up training on seed selection was provided to three farmer groups before paddy harvesting. The second phase of training comprised seed selection in the field, post-harvest activities and quality control methods. The farmers practised seed selection methods in the field and also learned simple germination and purity tests.

As a result of the training, a women farmers' group reported:

> From the training obtained through the SAGA programme on seed selection we have been able to select and produce good quality seeds for ourselves as well as for the market. Now, we don't have to depend upon others and go searching for quality seeds around the village.

ADCS members also provided feedback:

> For many years we had been selecting the seeds and saving it for the next season. However, the right way of seed selection from the very beginning results in significant change in quality of the seeds. The SAGA project provided us with good quality source seeds, which has helped us

to really take up the seed selection process very seriously. Due to the opening of market linkage, the farmers have been encouraged to grow the landraces.

Market Links

In strengthening markets for local landraces, our efforts contributed to the development of a new marketing network that integrates both formal and informal seed networks into a more structured, resilient configuration—in which notions of formal and informal become blurred (see Figure 4.10 compared with Figure 4.5).

**Figure 4.10
New Seed Marketing Channels**

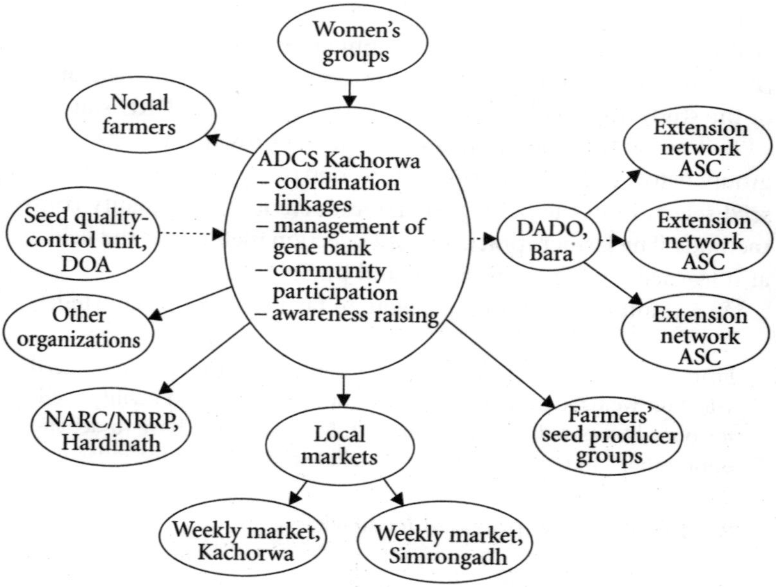

Notes: ASC = Agricultural Service Centre; ADCS = Agriculture Development and Conservation Society; DADO = District Agriculture Development Office; DOA = Department of Agriculture; NARC = Nepal Agricultural Research Council; NRRP = National Rice Research Programme.

A village-level workshop with representatives of agro-vets, the DADO, agriculture service centres and *golas* (local rice trading centres/shops) in and around the village was conducted to link the quality seed production programme with market outlets. Various stakeholders in the seed system were brought together into a common forum. Demands from the agro-vets and *golas* were noted at the forum.

ADCS, which is responsible for management of the local seed bank, plays a central role in the new marketing network. It carries out seed distribution, collection and quality control. Quality is monitored by ADCS in coordination with the quality control unit under the DOA. Marketing takes place through various formal and informal channels. The new marketing system integrates the formal seed network, which includes the agriculture service centres, agro-vets and the NRRP, with the informal networks, including the nodal farmers, seed producer group and *haat* bazaars.

Financial support for the project in the form of seed money to ADCS has allowed for the advertising and the establishment of seed stalls in the weekly markets to ensure the sustainability of the seed production programme. With this support and returns from initial seed transactions, the farmers of Kachorwa have begun producing seed from three more landraces that have high potential in the market. Farmers are very enthusiastic about these new opportunities.

Conclusions

Social and gender issues are important aspects to be considered in agriculture research, policy and extension activities, especially in a country like Nepal where the majority of the population depends on agriculture. Although various policies and development plans that mention social and gender issues or consider them to some degree have been in place for some time, the implementation part remains weak. The wide gap between policy formulation and implementation can only be narrowed by creating greater awareness of the vital importance of social and gender issues. Conceptual clarity is very important at all levels. Monitoring and assessment of how policies, programmes and projects have different effects on women and men, the poor and the rich, and various ethnic groups require strengthening throughout the agricultural research and development system.

Our case study aims to contribute to the way forward: how to systematically incorporate SAGA into concrete situations and crucial issues for farmers (that is, seeds). Our case study shows that culture and social structure strongly inform roles and responsibilities. Our analysis of the seed system in Kachorwa clearly shows that men across all wealth categories and ethnic groups are more involved than women in seed selection activities. Only a few women in the poor category are involved and only to a limited degree. Our research seems to show that the perception that women in South Asia are involved in seed selection may not be true in all cases.

Formal and informal systems are equally important in the flow of local and modern varieties. The formal system has a place in the flow of introduced varieties, whereas the informal system operating in the village has a significant place in the flow of both local and modern varieties. In the informal seed system a number of individuals play a significant role in the flow of genetic materials. These nodal farmers have their own network through which the flow of seeds and other planting material takes place. They tend to be richer farmers and men.

Farmers have developed their own methods for various farm activities. They have been practising seed selection for centuries and have considerable expertise. However, we found that their practices could be improved and we set out to do so by building capacity in quality control. Support to a community-based organization to carry out quality control at the field level will help ensure the sustainability of the seed production programme. Similarly, we expect that opening up new marketing opportunities for quality seeds of local landraces will provide incentives to farmers. 'Keeping landraces going' will also contribute to the conservation of local biodiversity.

The challenge of bridging the gap between improved local field experience and policy making remains. Our work aims to make a small contribution to this, but more efforts are required—by researchers, extensionists and policy makers alike.

References

Adamo, A. and A. Horvoka (1998). Guidelines for Integrating Gender Analysis into Biodiversity Research: Sustainable Use of Biodiversity Program Initiative. Ottawa: International Development Research Centre.

Brush, S. (1991). A Farmer Based Approach to Conserving Crop Germplasm. *Economic Botany*, 45: 153–61.

Central Bureau of Statistics (CBS). (1999). *Statistical Year Book of Nepal*. Kathmandu: National Planning Commission, His Majesty's Government.

Gauchan, D. (2000). Economic Valuation of Rice Landraces Diversity: A Case of Bara Ecosite, Terai, Nepal. In B. Sthapit, M. Upadhaya and A Subedi, eds, *A Scientific Basis of In Situ Conservation of Agro-biodiversity On-farm: Nepal's Contribution to the Global Project*, pp. 120–48 (NP Working Paper No. 1/99). Kathmandu: Nepal Agricultural Research Council/Local Initiatives for Biodiversity, Research and Development; Rome: International Plant Genetic Resources Institute.

Jarvis, D.I. (2000). Farmer Decision Making and Genetic Diversity: Linking Multidisciplinary Research to Implementation On-farm. In S. Brush, ed., *Genes in the Field: On Farm Conservation*, pp. 261–78. Ottawa: International Development Research Centre; Rome: International Plant Genetic Resources Institute; Boca Raton: Lewis Publishers.

Rana, R.B., P. Chaudhary, D. Gauchan, S.P. Khatiwada, B.R. Sthapit, A. Subedi, M.P. Upadhaya and D.I. Jarvis (2000). *In Situ Crop Conservation: Findings of Agro-ecological, Crop Diversity and Socio-economic Baseline Survey of Kachorwa Eco-site, Bara, Nepal* (NP working paper No. /2000). Kathmandu: Nepal Agricultural Research Council/Local Initiatives for Biodiversity, Research and Development; Rome: International Plant Genetic Resources Institute.

Subedi, A., D. Gauchan, R. Rana, S.N. Vaidya, P.R. Tiwari and P. Chaudhary (1999). Gender Roles and Decision-making in Rice Production and Seed Management: Experience from Different Agro-ecological conditions of Nepal. In B. Sthapit, M.P. Upadhaya and A. Subedi, eds, *Scientific Basis of In Situ Conservation of Agro-biodiversity On-farm: Nepal's Contribution to the Global Project*, pp. 38–48 (NP Working Paper No. 1/9). Kathmandu: Nepal Agricultural Research Council/Local Initiatives for Biodiversity, Research and Development; Rome: International Plant Genetic Resources Institute.

Subedi, A., P. Chaudhary, B. Baniya, R. Rana, R.K. Tiwari, D. Rijal, D. Jarvis and B. Sthapit (2001). Who Maintains Genetic Diversity and How? Policy Implications for Agrobiodiversity Management. In D. Gauchan, B.R. Sthapit and D.I. Jarvis, eds, *Agrobiodiversity Conservation On-farm: Nepal's Contribution to a Scientific Basis for National Policy Recommendations* (National Policy Workshop), pp. 24–26. Rome: International Plant Genetic Resources Institute.

Women for Development Division (WFDD) (2002). Annual Report. Kathmandu: Ministry of Agriculture and Cooperatives, His Majesty's Government.

5

Empowering Women Farmers and Strengthening the Local Seed System

Action research in Guangxi, China

Photo credit: Ronnie Vernooy (IDRC)

YICHING SONG AND LINXIU ZHANG WITH RONNIE VERNOOY

Changing Context and Challenges

China is experiencing major changes in its economy, paralleled by transformations in its society and large-scale environmental impacts. China's recent entry into the World Trade Organization (WTO) marked an important event in this period of change and is likely to accelerate the trends. The opening up of the country's economy to foreign enterprises and investment, and its closer integration into the global economy are undoubtedly having profound effects on the entire population. However, we expect that the reform will have different socioeconomic impacts on different groups of people depending on the sector in which they work, their location and factors such as gender and age. How the changes will affect the most vulnerable groups, such as small farmers, especially women farmers, and the implications for policy making have become an increasing concern to policy makers and researchers.

In this chapter we describe the results of an action-oriented research project entitled 'Empowering Women Farmers and Strengthening the Local Seed System: Action-Oriented Research on Open-Pollinated Variety Seed Production and Marketing'. The work was carried out by a social and gender analysis (SAGA) team in Guangxi as part of an ongoing research project coordinated by the Center for Chinese Agricultural Policy (CCAP). The larger project aims to improve farmers' livelihoods through the design, implementation and assessment of policies to support poor farmers based on a participatory, community-based natural resource management approach. Integrating a SAGA perspective is a core element of the work.

The Increasing Feminization of Agriculture

Recent studies in China have revealed that there has been an overall increase in migration from rural to urban areas, especially from the poorer areas, in terms of both number of households, number of people and duration of off-farm jobs (Song and Jiggins 2002; Song and Zhang 2004; UNDP 2003; Zuo and Song 2002). These studies also show that the out-migration of women is far below that of men in terms of its magnitude, time periods and increase over time. This means that more and more women are left alone at home and, as a result, have become more and more engaged in agriculture and household businesses.

Women are increasingly playing a key role in food production and farm management, while they continue to fulfil their usual household roles. How women (and men) experience this increasing feminization of agriculture, especially in the current context of increasing commercialization of agriculture, becomes an important question. It is evident that the feminization of agriculture has increased in the last decade in rural China, especially in the poor areas.

This trend was clearly shown in a recent CCAP study of 200 randomly selected farmer households in three poor provinces: Anhui, Qinghai and Guangxi (Table 5.1). There has been an overall increase in out-migration in terms of number of households affected, number of people migrating, and the length of time spent in off-farm jobs. For example, in 1997, migrants came from 84 households; this number increased to 101 in 1999 and jumped to 174 in 2001, with a further increase to 196 in 2003.

The same study also reported the overall trend in out-migration by sex between 1997 and 2003 (Figure 5.1). It showed that women's out-migration is limited and, although it has increased over this period, it is far smaller than men's.

Table 5.1
Changes in Out-migration in Anhui, Qinghai and Guangxi, 1997–2003

	1997	1999	2001	2003	Change (1997–2003)
No. of households with migrating member(s)	84	101	174	196	112
Average no. of migrating people/household	1.26	1.32	1.38	1.48	0.22
Average time away (months/year)	9.8	10.1	11.2	11.5	1.7

Source: CCAP; data collected in 2004.

Figure 5.1
Trends in Out-migration in Anhui, Qinghai and Guangxi by Sex

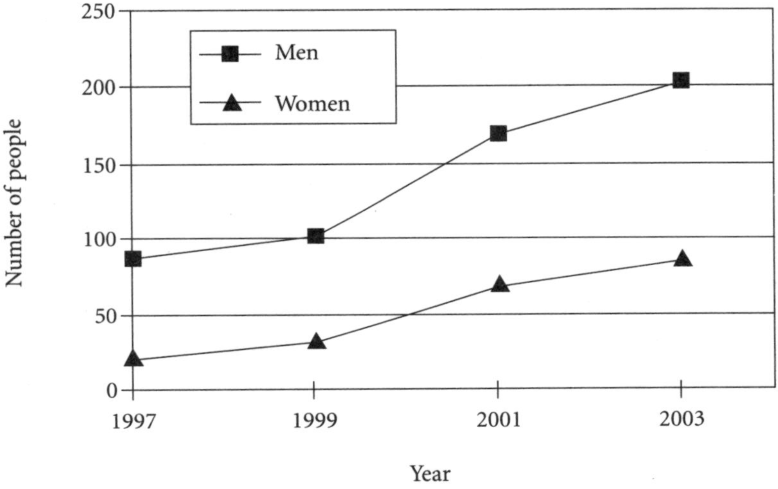

Source: CCAP; data collected in 2004.

The Increasing Commercialization of Agriculture and the Situation for Rural Women

Since the start of rural reform in 1978, the rural economy and agriculture in China have become more and more privatized and commercialized. The country's entry into the WTO is likely to bring about more modifications in agricultural production and commercialization, the rural economy at large and the (rural) employment situation all over China. In view of women's current inferior socioeconomic status and their increasing involvement in the agricultural labour force, some researchers argue that under the WTO, agriculture in China will become even more feminized and that many women are moving to more competitive labour-intensive agricultural activities (Song and Zhang 2004; UNDP 2003).

Without compensating measures and appropriate protection policies and regulations, we can expect a deterioration in rural women's economic, social and family status. This, in turn, will have a negative impact on the well-being of rural households, especially those that rely on agriculture for their livelihood in remote and resource-poor areas of China (Song and Zhang 2004; UNDP 2003).

Although China's economic growth has been impressive, poverty remains persistent in many remote rural areas, especially in the western regions. Approximately 30 million people still live below the absolute poverty line. They are mainly small subsistence farmers in resource-constrained remote western areas, which are agro-ecologically diverse, resource poor and risk prone. Women and women-headed households[1] represent a disproportionate share among the poor farmers in these remote regions. Women are becoming the main cultivators, food producers and on-farm income earners, but they also continue to play their traditional roles in household activities. Yet research has revealed that women are underrepresented in administrative and other roles, their importance is not recognized, and their specific needs, interests and expertise are largely neglected in agricultural and rural development (Song and Jiggins 2003; UNDP 2003; Zhang and Liu 2002). Evidence also shows that despite the significant role that women play in agricultural production, their access to basic resources and institutional services, such as credit, market information, training and extension services, has been limited. This, in turn, is having a negative impact on their livelihood security and that of their households (Song 2003; UNDP 2003; Zhang and Liu 2002; Zhu 2000).

Erosion of Genetic Diversity and Threat to Food Security

Careful balance of and trade-offs among economic growth, equity, poverty reduction and natural resource management are needed if long-term sustainable development is to be achieved. Erosion of genetic diversity of crops, especially food crops, is a very serious concern in China. For example, maize, which is the crop on which we focus our research, is now the most important food crop, the third most important food crop in China, and the staple of the poor in the south-western region. In its efforts to achieve national food security, the government of China has followed a modern, technology-oriented approach, relying predominantly on its formal seed system.

The development and distribution of modern varieties, mainly hybrids, for the three main staples, that is, rice, wheat and maize, has been the core task and the first priority in the formal system. Hybrid maize is now grown on approximately 80 per cent of the total production area in China, particularly the uniform and high-potential areas of the northeast and northern plain. The introduction of a market economy has resulted in an increasingly profit-driven seed production and supply system. Hybrid breeding and hybrid seed production have attracted more attention and investments than ever before. Conversely, a study done in Guangxi revealed that in the remote mountainous areas more than 80 per cent of the seed supply is from farmers' own seed systems, maintaining diversity for the interests and sustainable livelihoods of all farmers (Song 1998).

The genetic base for maize breeding in China has been dramatically reduced during the last decade. Although the national maize germplasm collection contains about 16,000 samples, five dominant hybrid varieties now cover 53 per cent of the total maize growing area of the country (Zhang et al. 2000). In Guangxi province the maize germplasm collection consists of about 2,700 samples, of which more than 1,700 are landraces from the region (Huang 2000). However, the use of these collected materials in breeding is very limited. Only three main hybrid breeding crosses are used, and all 14 hybrids bred out in the last 20 years share the same inbred line to different degrees (Huang 2000; Song 1998). Meanwhile, in other south-western provinces landraces in farmers' fields are degrading and disappearing as a result of the continuing spread of modern varieties.

Policy Context and Current Efforts

Fortunately, Chinese policy makers are increasingly aware of the links between biodiversity conservation, and sustainable development and poverty alleviation. The assessment of food security revealed that biodiversity loss is one of the new challenges facing China in its attempt to ensure food security for the future (Huang 2003). The government has realized that conservation and the sustainable use of biological resources are necessary if crop yields are to keep pace with the increasing population.

China—the most populated country in the world with the least amount of arable land per capita—has no choice but to keep food security high on its agenda. Although the government recognizes the phenomenon of the feminization of agriculture and knows that women are becoming the poorest among the poor, it has made little effort to do anything about this situation. The government has initiated some poverty alleviation and related programmes, but the cross-cutting nature of the gender issues and the crucial roles women are playing in biodiversity management, food security, poverty alleviation and rural development have still not been fully recognized and addressed.

Fortunately, several other organizations have started to tackle these issues in a more comprehensive way and with a multidisplinary approach, primarily through research projects. One of these efforts is a project, coordinated by CCAP a leading agricultural policy research institution that is part of the Chinese Academy of Sciences. CCAP project aims to identify technological and institutional options for developing more effective linkages and mutually beneficial partnerships between the formal and farmers' seed systems to enhance sustainable crop development and *in situ* or on-farm management of genetic resources, and to bring direct benefits to poor maize producers. At the same time, the project aims to strengthen women and men farmers' capacities to manage agrobiodiversity and improve their livelihoods (CCAP 1999, 2003). Given the fact that women are playing crucial roles but still have an inferior status in many rural areas, our work placed special emphasis on gender aspects and the empowerment of women. Gender issues have been central in all aspects of the work.

FOSTERING SYNERGY: OUR ONGOING RESEARCH

Our project 'Exploring the Potential for Crop Development and Biodiversity Enhancement: Fostering Synergy between the Formal and the Farmers' Seed Systems in China' started in January 2000, in Guangxi, with financial and technical support from the International Development Research Centre, Canada, and the Ford Foundation (see Figure 5.2) (CCAP 1999). The project involves the active participation of the main stakeholders, particularly farmers. It has established a strong multidisciplinary research team with main stakeholders from different levels.

Figure 5.2
Location of the Research Sites in Southern China

Source: Perry-Castañeda Library Map Collection.

Especially, it has built trust and established a local working network that includes enthusiastic and committed farmers and extensionists at the grassroots level and collaborative breeders at the provincial level (Vernooy 2001, 2003).

The project is being implemented by a team composed of women farmers, extensionists, breeders and researchers from different institutions and groups, with different disciplinary backgrounds and operating at different levels, from national down to the village. Five villages have been involved in the project as learning sites, as well as implementors (see Table 5.2).

Table 5.2
Basic Information about the Five Research Sites

Characteristic	Wenteng	Zicheng	Niantan	Zuron	Huaguang
Elevation (m above sea level)	52	660	460	580	620
Average annual precipitation (mm)	1,860	1,734	1,800	1,820	1,860
Agro-ecology	Open forest, hilly	Drought prone, mountainous	Drought prone	Open, flat valley	Drought prone, mountainous
Relative economic status	Better off	Very poor	Poor	Better off	Very poor
Distance from market (km)	2	10	6	3	12
Population	3,620	2,146	810	4,425	2,566
Inhabitants' average level of education (years of schooling)	7	4	6	7	5
Total labour force	2,839	1,500	450	2,200	1,560
Total no. of migrants	519	975	220	570	820
Female migrants	140	235	66	120	210
% of female-headed households	80	83	70	78	82
Average per capita land holding (ha)	1.62	0.94	0.8	0.79	0.95

Source: Adapted from Song (1998).

Our work builds on an impact study of CIMMYT (the International Maize and Wheat Improvement Centre) maize germplasm and poor farmers in south-west China. This study addressed the processes of technology development and diffusion by both the formal and the farmers' seed systems, and the impact of the introduced germplasm at different levels. One of the key findings was the systematic separation and conflicting operation of the formal and the farmers' seed systems, which resulted in poor adoption of formally bred modern varieties, an increasingly narrow genetic base for breeding, and a decrease and degradation of genetic biodiversity in farmers' fields (Song 1998, 2003). As our data show (Table 5.3) the number of varieties used in the research sites varies from three to five—a very small number.

Participatory plant breeding, seed selection and genetic resource management were the main methods we employed for improving maize to meet farmers' diverse needs in their heterogeneous environments and to enhance on-farm genetic biodiversity. Social learning was the other major method used. It focuses on characterizing the

Table 5.3
Key Characteristics of Maize Production at the Research Sites

Characteristic	Wenteng	Zicheng	Niantan	Zuron	Huaguang
Average maize yield (t/ha)	2.8	1.37	1.5	2.4	1.42
% of maize area irrigated	62	0	5	20	0
% of maize production area using chemical fertilizer	80	15	60	80	15
% of households using chemical fertilizer in maize production	100	31	80	98	35
% of households using purchased seed (hybrid) in maize production	85	0	15	70	5
Number of varieties used before research interventions	3	5	5	4	5
% of hybrid growing area	60	0	10	45	3
% of households engaged in maize production for commercial purposes	70	10	40	70	15

Source: Adapted from Song (1998).

formal and farmers' seed systems, eliciting farmers' knowledge, building on this information, and exploring appropriate methods and mechanisms to enhance the local organization of farmers and to empower farmers (Song and Vernooy 2003; Vernooy 2003).

Gender research and analysis was part of the initial baseline study. Subsequently, we examined a number of gender variables and issues, such as women's roles in agricultural production; their decision-making power in the household, community and society; their access to resources and services (credit, training); and their needs, interests and expertise in technology development and diffusion.

Tables 5.4 and 5.5 summarize the partial results of one of our village baseline surveys. Although women have become the main cultivators, farm managers and even factual household heads in many cases, in general, men still maintain their dominant role as decision makers (whether at home or when they are away in cities). It also shows that the traditional gender division of labour in rural China captured by the phrase 'the men till and the women weave' has been changing to 'the women till and the men work in industry'. However, the traditional ideology and old model, that is, 'men control the outside world, women

Table 5.4
Men and Women's Perceptions of Who Manages Household Resources and Activities

Resource/Activity	Men's perception (%)			Women's perception (%)			Difference (M–F) (%)		
	M	F	J	M	F	J	M	F	J
Land preparation	25	0	75	20	0	80	5	0	−5
Food crops	10	40	50	5	70	25	5	−30	25
Cash crops	30	20	50	25	45	30	5	−25	−20
Large livestock	65	10	25	45	30	25	20	−20	0
Pigs	0	82	18	15	65	20	−15	17	−2
Seeds	0	80	20	0	90	10	0	−10	10
Water collection	0	80	20	0	90	10	0	−10	10
Daily purchases	10	60	30	10	80	10	0	−20	20

Source: CCAP. The results are based on a baseline survey done in 2001 in the five project villages with a total of 40 respondents: 20 women and 20 men.
Note: M = men, F = women, J = joint.

Table 5.5
Men and Women's Perceptions of Who Makes Decisions about the Management of Resources and Activities

Resource/Activity	Men's perception (%)			Women's perception (%)			Difference (M–F) (%)		
	M	F	J	M	F	J	M	F	J
Land preparation	60	0	40	55	0	45	5	0	−5
Food crops	20	30	50	55	20	25	35	10	25
Cash crops	40	25	35	35	35	30	5	−10	5
Large livestock	65	10	25	70	0	30	−5	10	−5
Pigs	40	50	10	35	55	10	5	−5	0
Seeds	10	65	25	5	75	20	5	−10	5
Water collection	60	20	20	65	15	20	−5	5	0
Daily purchases	20	70	10	10	85	5	10	−15	5

Source: CCAP. The results are based on a baseline survey done in 2001 in the five project villages with a total number of 40 respondents: 20 women and 20 men.

Note: M = men, F = women, J = joint.

the inner', are still deeply embedded in people's behaviours and minds, including those of women themselves.

Table 5.4 compares men's and women's perceptions of their participation in resource management, and farm and household activities. In general, men and women think differently about the level of their own participation (if they were in agreement, the figures in the difference columns would all be zero, but that is true in only five cases). For some activities both men and women agree on job specialization by sex. For example, both men and women think that land preparation should be either men's or joint work. Also, seed and water collection should be mainly women's or joint work. In general, women believe they are participating more than men think they do. For example, among the eight listed activities, women think they had greater input into six of them than men thought (see the negative differences in the second last column of Table 5.4). For only one activity (raising pigs) was men's perception of women's participation greater than women's perception. In contrast, in half of the activities men thought that they had more input than women thought they did.

Other differences exist concerning certain activities. For example, most men think that producing food crops and cash crops should be mainly joint work, while most women think that this should be mainly women's work. However, there is a certain broader consistency on certain perceptions. For example, most men (75 per cent) and women (80 per cent) think that land preparation should be joint work. Also, the majority of both men and women think that raising pigs, producing seeds, collecting water and purchasing daily necessities should be women's work.

Table 5.5 compares men's and women's perceptions of who makes decisions about resource and activity management. In general, there are smaller differences between perceptions about decision making than there are about participation (compare absolute numbers in the difference columns in Tables 5.4 and 5.5). Both men and women think that men should be the major decision makers when it comes to preparing land, raising large livestock and collecting water. Both sexes also think that women should be the main decision makers in such activities as raising pigs, producing seeds and purchasing daily necessities. In certain activities both men and women think that men make the decisions but do not participate in the activity. For example, more than 60 per cent of both men and women said that men make decisions about water collection, but none of the participants thought that men actually carry out this task. A similar observation can be made about seed production. The largest difference between men and women in decision making is over food crops. Although 50 per cent of men think that decisions should be made jointly, 55 per cent of women think that decisions are actually made by men.

Given the nature of our project, specific attention was paid to analysis of the needs and interests of women farmers, who are the main cultivators and seed selectors in the area. Gender analysis and comparisons in variety selection criteria were used in the technical aspect of the project. We found that both women and men are mainly concerned with drought-resistant varieties—the key issue in poor, rain-fed areas. However, women were interested in various other criteria, which they ranked highly: production of seed that can be collected by the farmers and used the following year, low fertilizer requirements and good cooking quality (Table 5.6). These results reflect women's specific interests and needs

Table 5.6
Comparison of Variety Selection Criteria between
Women and Men Farmers in Guangxi Villages

	Frequency of selection (%)	
Criterion	Women (n = 20)	Men (n = 20)
Drought resistance	100	100
Lodging resistance	90	83
High yield	80	83
Seed for following year	80	50
Grain colour	70	50
Cooking quality	50	33
Plant shape/intercropping	50	83
Low fertilizer rate	40	33
Maturing time	40	33
Plant height	30	33
Rate of damage resistance	30	33
Disease resistance	20	33
Insect resistance	20	33
Growth cycle	10	50

Source: CCAP. Data collected in the project villages in 2001. Each person could select more than one criterion.

as producers as well as housewives, and should be considered in the technology development process.

INTEGRATING SOCIAL AND GENDER ANALYSIS TO EMPOWER WOMEN FARMERS AND STRENGTHEN THE LOCAL SEED SYSTEM

The geographic and economic dimensions of the trend toward the feminization of agriculture indicate that the poorer areas have the largest proportion of women in agriculture and the lower the household income, the greater the proportion of female-headed households. At same time, the public research and extension systems in China have been 'designed for men' and are predominantly male-staffed. To them, 'farmers are farmers', and SAGA is irrelevant. In addition, it is the village men (returning from town on leave from their work) who attend extension meetings and training sessions, even though they are no longer active farmers.

Women farmers and their groups have been key research partners in our project from the beginning. Despite this, we felt that more effort and action-oriented activities were needed to further integrate gender perspectives into the project process while also building the capacity of the team and empowering women farmers. This becomes essential, not optional, for the project and for a sustainable rural and agricultural development process in China, as women play multiple and crucial roles in biodiversity enhancement, crop improvement and rural livelihood security.

Action Objectives and Research Questions: The Proposed Agenda

Our key research question was: What kind of action-oriented activities can empower women farmers? Based on a series of discussions within the team and with the women farmers in the project villages, production and commercialization of open-pollinated seed varieties (OPVs) were identified as initial actions to add value to women farmers' traditional knowledge and the local process. Through women farmers' involvement in the market system, we aimed to empower them economically and politically. Specifically, we set out to encourage and enable women farmers to be involved in seed marketing; develop the capacities of women farmers in seed production and marketing; and enhance the local seed system and indigenous networks. The major research questions were:

1. How can the SAGA research team and project appropriately empower women farmers and enhance the farmers' seed system? What are the right activities and appropriate methods?
2. What internal and external factors can affect the empowerment and enhancement process?
3. What are the implications for policy making?

Participatory Planning

First, we discussed and shared the results of the first international SAGA workshop in Beijing (see Chapter 1) with the team, including women farmers, local extensionists and breeders, during our field and village visits. Because the initial idea of an SAGA action plan was raised by women farmers and supported by the other main social actors in the project, we proceeded directly to a discussion of our specific objectives and related methods and activities. We discussed the selection of two villages for in-depth work: Wenteng and Niantan, the two participatory plant breeding trial villages. Based on these discussions we refined our action plan.

On approval of the plan, a workshop was organized at the beginning of July 2002 with women farmers, local extensionists and breeders from the two selected villages. The workshop helped us prepare a work schedule and assign tasks to team members. The women farmers were enthusiastic and active in the planning process. They reiterated that they saw action on OPV seed production and marketing as a crucial entry point for their empowerment, economically and politically. They were proud to be at the forefront of this initiative, and made some innovative suggestions for improving the plan. For example, because women farmers work as a group in the seed production villages for both social and technical reasons, they suggested that the first seed production trials should target four OPVs rather than two as originally planned, to provide more options for different users. Also, they suggested that a farmer seed fair in one of the villages would open the window to more sources of varieties, from both the farmer and formal seed systems.

Fieldwork

Given that the season for the second maize crop starts in July, immediately after the planning workshop the team went to the two selected villages. Varieties were chosen and trial fields were identified by the women farmers with assistance from the local extensionists and breeders. Four varieties—961, 963 (from the CIMMYT population), farmer-improved Tuxpeno and a newly bred variety known as Mexico 1— were selected for seed production and were sown in farmers' fields in July. Good quality seeds for these varieties were provided by the Guangxi Maize Research Institute and by a woman farmer breeder in the team who has 20 years of experience in maize improvement (for more details, see Song 1998). Training in seed production was carried out in the field by experienced women farmer breeders, local extensionists and formal breeders.

After the sowing season some group discussions took place concerning the marketing of the seeds produced. Three major market channels and users were identified and it was agreed that they would be the targets for seed dissemination: the existing farmer-to-farmer exchange network, farmer to local seed market, and farmer to the external market. For the existing farmer-to-farmer exchange channel, it was decided to 'just keep it as it is' and observe and document the dissemination and diffusion process. A farmer seed fair was organized at the beginning of 2003, and this proved to be an innovation in China and a good way to encourage and enable women farmers to enter the market and further strengthen, formalize and legalize the existing local farmers' seed marketing process. Some efforts were made to link the women farmers with the external seed market to add value to their product and empower them in the process. Some external market research was begun in three potential regions: Ningxia, Jiansu and Gansu provinces. Two of the provinces have already imported some seeds from the Guangxi farmers for field adaptation trials.

Results

Enhancing Women's Organizational Skills

The SAGA action plan was initiated by women farmers and implemented mainly by women farmer groups in collaboration with grassroots extensionists. New links have been formed allowing farmers collectively to channel ideas and demands into the formal extension and research system. In the absence of institutional support, some women farmers support each other and have organized, for example, to exchange information, skills and experience; to share labour; and to set up and manage micro-credit. The capabilities and bargaining power of the women in these groups have been strengthened. Women farmer groups also continue to be actively involved in the design and implementation of participatory plant breeding (PPB) field experiments. The groups differ from each other in terms of purpose and capacities, as they are influenced by their environmental and socio-economic conditions. However, they are all enthusiastic and play important roles.

The SAGA work has further enhanced the capacity of the women's groups for self-organization and management. They took part in training in management, seed production and marketing; their access to seed market information has increased; and they are now becoming directly involved in the formal seed market. Moreover, these women have become more and more dynamic in their households and communities. They are playing a leading role in production and income-generating activities in terms of participation and decision making.

Strengthening the Linkage between Women's Groups and the Extension System at the Grassroots Level

To facilitate the exchange of farmers' and formal knowledge and to strengthen links between the two seed systems, we identified the grassroots extension system (including the township extension stations and their village farmer technicians) as another local network to work with and strengthen. In the transition from a planned to a market economy, the role of the public extension system is changing from one of serving only the state to serving the farmers as well as the state. This represents a major shift towards a more client-oriented and demand-driven system. The township extension stations are operating under very difficult conditions and are facing financial constraints. Yet they are the only source of formal technical assistance available and accessible to farmers. They have formally trained personnel, who are almost all local people from farming communities. These local 'experts' are familiar with the farming systems and are more enthusiastic than outside professionals about using a participatory approach in working with farmers.

Through selected project activities, staff at these stations are working with women farmer groups, for example, through PPB and participatory varietal selection (PVS) experiments. Good collaboration among the women farmers, township extensionists and village technicians has been established in all five villages. This collaboration, which is highly appreciated by farmers, has become the institutional basis for the design and implementation of a new policy experiment in formal collaboration with the Ministry of Agriculture. The experiment is part of the country's agricultural extension reform process currently under way. In our case, it will build on the work done to date and reinforce the SAGA efforts.

Adding Value to Farmers' Varieties, PPB Products and Local Process by Involving Women Farmers in OPV Seed Production and Marketing

The four varieties selected for seed production have all produced very good harvests, especially the PPB variety, Mexico 1. Its yield is 15 to 20 per cent higher than currently cultivated OPVs and has a number of characteristics that farmers, especially women farmers, prefer—drought tolerance, lodging resistance and good cooking quality. The other three varieties (961, 963 and Tuxpeno) also produced good harvests and met farmers' specific needs and interests, but their characteristics differ. The seeds of all four varieties have been disseminated through the existing farmer-to-farmer exchange network.

Some seed has also been distributed through local seed markets, such as the free farmers' market and the farmers' seed fair. Mexico 1 has become popular with farmers and in the local market and was nominated for the national variety test in 18 provinces in 2003. Results of these experiments are now being analysed.

Farmers' Seed Fair in Guangxi: An Innovation for Enhancing Recognition and Exchange of Farmers' Knowledge and Resources

The team suggested that a seed fair would be a good way to encourage and enable women farmers (and men as well) to become involved in seed marketing, to learn more about maize and other varieties and their characteristics, and to explore how best to legalize existing local seed marketing systems and exchanges. Following harvest of the seeds, the first farmers' seed fair in Guangxi (and maybe in China) was organized; it was considered a success and a significant event (see Box 5.1).

Box 5.1 The Seed Fair in Guozhai

It is market day in Guozhai township, Guangxi province, and there is excitement on the streets because the fair has come to town. No ordinary fair this, it is an agricultural biodiversity fair, the first such event ever seen in the province—or very likely anywhere in China.

Hundreds of local farmers crowd the section of the main street where the fair has been set up to view the diversity of crops and seeds on display. Others are there, too—township officials, merchants, curious children, important visitors from Beijing, even a local television crew filming the action. What they see is a rich diversity of crops—38 crops and 107 varieties. Some are rare and unique to the area, such as black wax maize and mountain lily. They include 31 varieties of maize, 17 of beans, 16 of vegetables, 14 of cereals, and 8 root crops. As well, there are traditional herbs, spices and medicinal plants—almost all of them landraces.

This cornucopia is displayed in booths just as at the regular market, and at each booth there is a neatly printed card giving details of the type and origin of the items laid out. Proud farmers and some researchers and extension agents staff the booths, happy to talk about their produce with anyone who asks—and many do.

The fair-goers are impressed. 'I didn't know it was possible to grow so many varieties of crops here,' says one. An older farmer examines the maize on display and shakes her head in wonder. 'I haven't seen these seeds since the 1960s,' she says. 'There are seven maize varieties here that I've never seen before,' responds her companion.

Throughout the day the participants—farmers, researchers and officials—exchange many opinions, ideas, experiences and, of course, seeds. In the afternoon a committee that has been assessing the wealth of diversity at the fair awards prizes to the most outstanding displays. The first agricultural biodiversity fair in Guangxi province is attended by more than 2,000 people. It is a huge success—and it will not be the last such event. (Vernooy and Song 2003)

Emerging Policy Issues

Reflecting on our research questions, we can summarize our major findings and preliminary conclusions. First, our SAGA action research has shown that it is crucial to add value to farmers' traditional knowledge and their resources, such as landraces and self-improved varieties. Our case study illustrates that marginalized women farmers and their knowledge can be easily recognized, protected and strengthened through a collaborative process in which farmers and the researchers, extensionists and policy makers in the formal system work together on a more equal basis.

Despite the progress made, more appropriate and supportive marketing, institutional and legal mechanisms and policies are needed to involve more farmers and improve the quality of their participation. Our case also illustrates that SAGA is essential, not optional, for the formulation of responsive and gender-sensitive policies, regulations, and related implementation and management strategies to avoid further marginalization and biases in the mainstreaming process.

The women's groups in our case study show that it is important to protect farmers' abilities to organize, manage and empower themselves. Strengthening women's groups (or other farmers' groups) and the communities of which they are part can be the first step in self-organization and autonomous capacity building. Farmers' seed fairs seem to be a promising innovation. In a short time, they became very popular, enhancing the recognition and exchange of farmers' knowledge and genetic resources. In addition, the fairs can enhance local seed systems and indigenous networks by adding value to farmers' knowledge and local processes, economically, culturally and socially. Collaboration between women's groups and grassroots extensionists is a way to enhance the access of women farmers to institutional support and their links to the formal system.

Although progress has been made, we continue to face many challenges. A number of policy issues have emerged. Will entry into the WTO favour or constrain women farmers' seed enterprises? How will new rural development policies, such as rural extension reform, take into account and affect the differential views, needs and interests of rural women and men? Will the new policies allow women to take collective action or further contribute to the feminization of agriculture and poverty? We have set the goal for ourselves to continue working on the integration of SAGA in four policy areas. We describe these briefly (CCAP 2003).

Provide Space and Support for Farmers, Especially Women, to Establish and Operate Viable Seed Enterprises to Improve Incomes and Livelihoods more Broadly

The new seed policies and laws allow private seed production for all kinds of seeds, including OPVs. However, right now the large seed companies (both private and public) are only interested in the production and marketing of hybrids. Their main interest is in making money; genetically modified seeds are now also being marketed. So far, little effort has been made to serve the poor and marginal upland farmers who rely on local landraces and OPVs. There are at least two challenges: (*a*) to provide support and incentives for small farmers to enter into the seed market (in particular for the OPVs and the new PPB varieties); and (*b*) to harmonize the new seed policies and laws with other broader agricultural policies and laws affecting seed production and commercialization, for example, policies on rural extension and WTO-related changes.

Clarify the Questions of Recognition, Access and Benefit Sharing Related to the Development of New Varieties through PPB

PPB varieties are the result of the efforts of many people: farmers, plant breeders, other researchers, extensionists and maybe others such as donors and policy makers. This raises many questions—about the proper recognition of the contributions to the new variety (ideas, knowledge, skills, time, energy, money and other resources); about adequate access to and use of the new varieties; about the commercial and non-commercial benefits that the new varieties bring; and about sharing these benefits fairly. Right now there are policies and laws regarding plant breeders' rights, but none that recognize the rights of small producers. In some countries ideas are being put forward and action is being taken to establish so-called farmers' rights. Is there a need for new policies and legislation that deal with PPB varieties in China in terms of recognition, access and benefit sharing?

Fortunately, Chinese policy makers are increasingly aware of the links between (*in situ*) biodiversity conservation, and sustainable development and poverty alleviation. The leading maize breeder in the country is now convinced that the *in situ* conservation of landraces is crucial for the national programme to broaden the genetic base for maize. This is promising, but the challenge is to turn this new perspective into policies and laws. Not only for maize, but also for other crops on which poor farmers rely.

Reorient Rural Extension Services toward Serving Small Farmers and Responding to Client Demand

At the moment the extension service in many provinces of China, including Guangxi, has 'collapsed'—a fact that is actually being recognized by the higher level authorities in charge of rural extension reform. These authorities are looking for alternatives. We think that extension services must be demand driven and responsive to poor farmers' needs in terms of providing information and linking local interests and views to government service providers, to research (such as plant breeding) and to education. The challenge now is to develop and test practical alternative service provision mechanisms. Our project has already been experimenting with a participatory extension approach by integrating grassroots extensionists with researchers, breeders and policy makers. More experimentation is required, for example, concerning finance, incentives, costs and benefits, and institutional set-up of the services at the township and county levels.

NOTE

1. Household head here means the de facto head, who is responsible for the daily management of the farm and the rural household most of the time. Despite the fact that most households are now headed by women, their husbands continue to be formally (and legally) registered as household heads in most cases.

REFERENCES

Center for Chinese Agricultural Policy (CCAP) (1999). Exploring the Potential for Crop Development and Biodiversity Enhancement: Fostering Synergy between the Formal and the Farmers' Seed Systems in China. Research Proposal, CCAP, Beijing.

────── (2003). Rural Livelihood Security and Policy Changes: Enhancing Community-based Crop Development, Natural Resource Management and Farmer Empowerment in Guangxi, SW-China. Research proposal, CCAP, Beijing.

Huang, J. (2003). Food Security in China Re-considered. Report submitted to the Chinese government, CCAP, Beijing.

Huang, K. (2000). Analysis on the Development of Hybrid Breeding in Guangxi and its Relationship with the Genetic Base and Heterotic Patterns. *Journal of Guangxi Agricultural Science*, November, pp. 36–39.

Song, Y. (1998). 'New' Seed in 'Old' China: Impact of CIMMYT Collaborative Programme on Maize Breeding in South-western China. PhD thesis, Wageningen University and Research Centre, Wageningen.

────── (2003). Formal System and Farmers' System: The Impact of Maize Germplasm in Southwest China. In S. Mathur and D. Pachico, eds, *Agricultural Research and Poverty Reduction: Some Issues and Evidence*, pp. 173–89. Cali: Centro Internacional de Agricultura Tropical.

Song, Y. and J. Jiggins (2002). The Feminization of Agriculture and the Implication for Maize Development in China. *LEISA*, 18(4): 6–9.

────── (2003). Women and Maize Breeding: The Development of New Seed Systems in a Marginal Area of South-west China. In P.L. Howard, ed., *Women and Plants: Gender Relations in Biodiversity Management and Conservation*, pp. 273–88. London and New York: Zed Books.

Song, Y. and L. Zhang (2004). Gender Assessment Report: Impacts of IFAD's Commitment to Women in China 1995–2003, and Insights for Gender Mainstreaming. Working Report, International Fund for Agricultural Development, Beijing.

Song, Y. and R. Vernooy (2003). Participatory Maize and Livelihood Improvement in Southwest China. *Agricultural Research and Extension Network Newsletter*, 48 (July): 8.

United Nations Development Programme (UNDP) (2003). Overall Report on China's Accession to WTO: Challenges for Women in the Agricultural and Industrial Sector. Collaborative research report, UNDP, United Nations Development Fund for Women, All-China Women's Federation, National Development and Reform Commission and Center for Chinese Agricultural Policy, Beijing.

Vernooy, R. (2001). Harvesting Together: The International Development Research Centre's Support for Agrobiodiversity Research. Working report, International Development Research Centre, Ottawa.

────── (2003). *Seeds that Give: Participatory Plant Breeding*. Ottawa: International Development Research Centre. Available at http://www.idrc.ca/seeds (viewed 12 February 2005).

Vernooy, R. and Y. Song (2003). Celebrating Diversity in China. *LEISA*, 19(3): 36.

Zhang, L. and C. Liu (2002). Gender and Equity Issues in Land Tenure Arrangements in China. Research paper, Center for Chinese Agricultural Policy, Beijing.

Zhang Shihuang, Peng Zebin and Li Xinhai (2000). Heterosis and Germplasm Enhancement, Improvement and Development of Maize. *China Agricultural Science*, 33: 34–39.

Zhu Ling (2000). Gender and Equity Issues in Land Allocation. *Economic Research*, 9: 39–46.

Zuo, J. and Y. Song (2002). *Women's Experiences with 'Feminization of Agriculture': Insight from Two Village Case Studies in Southwest China*. Beijing: Qinghua University (in Chinese).

6

Creating Opportunities for Change

Strengthening the social capital of women and the poor in upland communities in Hue, Viet Nam

Photo credit: Ronnie Vernooy (IDRC)

HOANG THI SEN AND LE VAN AN

The Region and the Issues

Agriculture continues to be one of the most important sectors in Viet Nam's economy. It provides a livelihood for over 70 per cent of the population and contributes to the national economy through the export of rice, coffee, rubber and other agricultural products. Since the implementation of the *doi moi* or reform policy, Viet Nam's economy has changed significantly; from 1990 to 1997 the average annual growth rate was about 8 per cent. Since 1998, although the country was somewhat affected by the 1997 financial crisis in Asia, the gross domestic product (GDP) has still increased by about 6.5 per cent annually. Per capita income rose from USD 220 in 1994 to USD 320 in 1997, and now exceeds USD 400.

Despite the progress at the macroeconomic level, poverty persists and remains a challenge in the development of the country. Of the approximately 80 million people in the country, 29.6 million (37 per cent) have incomes below the national poverty line and are considered poor. The poorest people in Viet Nam are rural farmers living and working in the mountainous areas. Many factors, such as increasing population pressure, cultural values that differ between ethnic minority groups, difficulties in cross-cultural communication, and degradation of natural resources, make it difficult for those farmers to attain sustainable livelihoods.

Approximately three-quarters of Viet Nam's natural land area is classified as uplands and is home to 25 million people who belong to 50 of the 54 different ethnic groups in the country. In these areas low agricultural productivity, widespread poverty, changing migration patterns and the marginalization of Viet Nam's diverse ethnic minorities are indicators of the broader social and environmental challenges confronting the country's future development. In the uplands the main issue is poverty alleviation coupled with conservation of natural resources. Sustainable management of soil, water, forests and other natural resources is critical, not only for the local people but also for the nation as a whole.

In response to this critical need and the problems facing many communes and villagers in Thua Thien Hue province and other parts of central Viet Nam, the Community-based Upland Natural Resources

Management (CBUNRM) project was developed at the University of Hue, with support from IDRC and the Ford Foundation. The project is being implemented by a research team from the Hue University of Agriculture and Forestry and other partners. Hong Ha commune was selected as the research site on the basis of its social, economic and natural conditions, which are typical and represent the upland situation for many communes in central Viet Nam. Through this project we wanted to gain a better understanding of the links between poverty, policies and resource degradation, and to test alternatives for improving agricultural production systems and building human, social and natural capital.

For example, some recent changes in forest policies are encouraging. Local authorities have been meeting with us and with villagers to discuss various possible joint management arrangements or agreements. One of our aims is to 'make policies work for the poor'. This requires involving different stakeholders at district, provincial and even national levels. The overall goals of the CBUNRM project are to develop materially better livelihoods for the poor in upland communities; to advance human resource capacities of various groups, including community members; and to make policies and programmes perform for the poor. We believe that one alternative is to develop and establish mechanisms that encourage the local people to generate locally adapted strategies for sustainable development. Such a human capacity building approach is at the heart of our work.

We are using a variety of communication tools and participatory methods, including appraisal, monitoring and evaluation. Over time, we have seen the degree of farmer participation increase and results being more readily and widely adopted. We have also learned that farmers adopt simple techniques more quickly. As farmers gained confidence in the researchers, they accepted suggestions more readily and expressed their own ideas more openly. Among the ethnic minorities in the uplands, conducting research with the participation of some farmers draws the interest of other farmers. Action research, in which farmers can see what is happening, increases their understanding and invites their opinions. This goes beyond just making recommendations. Using a participatory approach has helped us to direct attention to the role and needs of women. While men talk, it is often the women who do most of the work in the fields. Farmer participation is also useful in evaluating research results. Group meetings and on-farm workshops give farmers an opportunity for self-assessment.

A second pillar of our method is a community-based perspective. This approach provides an opportunity for building capacity at the community level: everyone benefits and confidence in local leadership increases. Strong and experienced leaders, in turn, can strongly influence the participation of all members. Encouragement of social activities helps build a sense of community and close relations between the researchers and the community. Strong leadership and a united community are important preconditions for implementing co-management of natural resources by the community and the government agencies.

Rationale for the Integration of Social and Gender Analysis into our CBUNRM Research

Women around the world have less access to information, training and education. A global survey shows that women receive only 5 per cent of all agricultural extension services worldwide (FAO 1997). A survey in the Mekong Delta showed that 72 per cent of women's labour was in agricultural work; 82 per cent were responsible for housework. Their educational level was low and they did not receive much technical guidance (Luat and Son 1992). Compared with lowland women, those in the uplands have even less access to information and technology. There are many reasons for this. Women spend long hours working in the field and at home, their education levels are generally lower than those of men, and traditional perceptions about men's and women's roles hinder women's mobility and involvement in activities outside the home or farm.

Our study in Hong Ha (Hoang 2000) showed that women spent 10 times as much time doing housework than men. They also spend more time working in the field than men, but have a lesser role in decision making and less control over resources.

Development programmes designed and implemented by the government and rural communities do not consider social and gender factors. Among the many development programmes executed in the uplands, only a few pay attention to the role of women and their ability to contribute to community activities. In general, decisions about development are mainly made by men. To address this situation, our CBUNRM research project made capacity building for women and the poor one of the main targets. We think that building assets for the upland poor is a key approach to sustainable development.

The social and gender analysis (SAGA) research that we describe in this chapter aimed to strengthen the capacity of research team to integrate SAGA into CBUNRM research; enhance awareness of and sensitization to SAGA among researchers and development workers in their work with farmers; and help women and the poor solve problems by themselves to alleviate poverty and contribute to the sustainable development of their upland communities.

We formulated the following research questions:

1. How can the research team strengthen its approach to working with the poor and women, and improve their capacities?
2. How can researchers and district officers increase their awareness of the poor and women in the community, and do better diagnosis and planning, strengthen local organizational capacities in technology development, and improve extension services?
3. What are the constraints on such capacity building efforts?
4. What are the impacts of such capacity building on the livelihood of the upland poor and women?

THE RESEARCH SITE, TEAM BUILDING AND FIELDWORK

The Research Site

Our work was carried out in the Hong Ha and Huong Nguyen communes of A Luoi district, Thua Thien Hue province, Viet Nam. Hong Ha and Huong Nguyen are two of the 16 poorest communes in A Luoi district and among the approximately 1,200 designated 'poorest communes' in the country according to national poverty criteria. Hong Ha was the initial research site; Huong Nguyen is a new site, where lessons learned from Hong Ha will be evaluated and disseminated in cooperation with various agencies, particularly the provincial

Strengthening Social Capital of Women and the Poor: Viet Nam 161

Figure 6.1
The Viet Nam Research Site

Source: Perry-Castañeda Library Map Collection.

Department of Agriculture and Rural Development (DARD) (see Figure 6.1). The total land area of the two communes is about 47,000 ha. Most of this land is owned and managed by the state. In Hong Ha most of the land in and around the commune is now under 'watershed protection and management' by the Bo River Watershed Department (a government agency). In practice, this means that local people only have access to and control of about 1 per cent of the total land area for agriculture. The portion of agricultural land is very small and is used mostly to cultivate rice and cassava (Table 6.1).

Hong Ha and Huong Nguyen communes are the home of the ethnic minority groups Ka Tu and Ta Oi, who live in the uplands of Thua Thien Hue province (Table 6.2).

Table 6.1
Land Use in Hong Ha and Huong Nguyen Communes

Land use	Hong Ha (ha)	Huong Nguyen (ha)
Total land area	14,100	32,590
Agricultural land	180	98
Paddy rice	20	19
Upland rice	50	27
Cassava	70	15
Other agricultural use	40	37
Forest land	11,000	18,914
Natural forest	10,200	18,800
Plantations	765	114
Unused land	2,950	13,419

Source: CBUNRM project fieldwork data, 2002.

Table 6.2
Population of Hong Ha and Huong Nguyen Communes

Population group	Hong Ha	Huong Nguyen
Total population	1,266	1,050
Ka Tu	609	998
Ta Oi (including Paco and Pahy groups)	520	3
Kinh (lowlanders)	137	49

Source: CBUNRM fieldwork data, 2002.

Local people depend largely on agriculture; most of what they produce is for home consumption. The traditional practice has been slash-and-burn farming; however, as land has become scarce and with the introduction of a government policy against this method, sedentary farming is now the common practice. Farmers are facing many problems as they try to convert to sedentary farming. Among the main ones are unfamiliarity with new production techniques, poor access to technologies and other inputs, and lack of knowledge about soil management practices, with resulting low productivity.

It is important to note that Huong Nguyen has been newly settled under a government programme, with only about 10 years of local history. Originally, Huong Nguyen villagers lived in a remote, inaccessible mountain valley with a very rich forest area near the Viet Nam–Laos border. When they were resettled, they were allocated unproductive *imperata* grasslands or wastelands that needed to be converted to agricultural fields, and this is a long and difficult process. They also received some farming tools and food to help them begin their new lives. So, unlike Hong Ha villagers, who moved back to their own homelands after the war, Huong Nguyen villagers were forced to resettle. Table 6.3 shows the intra-commune division of wealth.

Table 6.3
Wealth Ranking of Commune Households*

Socioeconomic group	Hong Ha	Huong Nguyen
All households	243	185
Better off	1	4
Middle	123	21
Poor	30	65
Very poor	92	95

Source: CBUNRM fieldwork data, 2002.

Note: *According to national wealth indicators developed by the Ministry of Labour and Social Affairs (note: 15,000 Vietnamese dong [VND] = 1 United States dollar [USD]): better off = monthly income over 100,000 VND/person; middle = monthly income 80,000–100,000 VND/person; poor = monthly income 55,000–80,000 VND/person; very poor = monthly income less than 55,000 VND/person.

Team Formation and Fieldwork

To carry out the research in these communes as part of our broader CBUNRM action research agenda, we established a SAGA team. It included lecturers from the Hue University of Agriculture and Forestry with various academic backgrounds (animal husbandry, agronomy, horticulture, forestry and aquaculture); lecturers from the Economic University who specialized in agricultural economics and policy analysis; and lecturers with sociology training from the Science University. This team developed and implemented a framework for integrating SAGA into the CBUNRM work in close collaboration with development workers and leaders in the communes. A research agenda was developed aimed at improving the research capacity in social and gender issues, in particular to better understand and work with the very poor. We also aimed to strengthen our ability to understand and cooperate with development workers from different line departments and other groups.

We found it difficult to get men and women farmers and other stakeholders to participate in the whole process of action research as they are more used to a top-down approach. Traditionally, researchers take the lead at all stages—identifying problems, implementing and evaluating solutions—and the level of farmers' participation is low or nil. We learned that a good participatory approach requires a variety of quantitative and qualitative methods. Patience is also required to allow time for learning. Meeting the practical needs of the farmers and increasing their confidence are key elements of any capacity building process.

A series of training sessions was organized to introduce and discuss approaches, concepts and ideas about SAGA. These sessions built on previous knowledge of participatory research concepts and skills in techniques as described earlier. Local leaders, researchers and development workers took part in the training sessions. At the commune level the participants included commune leaders, chairpersons of local organizations, such as the women's union and the farmers' association, and leaders of hamlets. Experts in social sciences, policy analysis and gender were invited to facilitate the training.

Our social and gender research methods included participant observation, participatory rural appraisal, formal survey techniques, the formation of farmer interest groups, and participatory monitoring and evaluation. We briefly discuss some of these.

Farmers' needs and abilities varied. We worked with women and the poor to identify what they need and what they can do. With our support, farmers then formed learning groups, in which they shared information about their practices and interests in terms of seeking improvements. These groups were built around agricultural commodities: rice, cassava, home gardens, fish and pigs. Each commodity served as an entry point to the larger farming system, so group interests were always broader than their main commodity. Several types of interest groups were formed based on the premise that farmers have different interests, aspirations and needs.

We carried out a survey and specific participatory rural appraisals among about 15 per cent of the households. Sixty families representing the various wealth categories were interviewed about a wide range of issues concerning their livelihoods. We also interviewed leaders of the commune, women union representatives and leaders, farmer association leaders and leaders of hamlets. In-depth interviews were carried out with case study families based on the main research interests or topics.

Throughout the research cycle we used participatory monitoring and evaluation to keep track of our efforts and assess both results and the participatory process. Data were recorded and shared with the various stakeholders. Indicators were developed to measure the access of women and the poor to technologies and other inputs (for example, knowledge). Two workshops were conducted in the villages with local people, researchers and extension workers to discuss the research results.

CONVENTIONAL EXTENSION PRACTICES UNDER THE LOOP

Traditional Extension Tools for Development Projects in Upland Communities

Training, study visits and local television programmes on extension issues are the main sources of information and knowledge for people in upland communities. Our survey showed that training sessions and study visits focus on agricultural production technology, such as animal production, crop production and forestry. Crucial information about credit, health care and education is almost absent. (Some information about family planning is provided by the women's union.) Access to training sessions and study visits differs by wealth category: the poor have fewer opportunities than the middle and better-off families (Table 6.4).

Both the men and women interviewed indicated that men attend technical training sessions more often than women, although women spend more time than men in most production activities. Opportunities also vary between families. Women in poor households have less opportunity to attend training than men (Figure 6.2) and, in general, women attend training sessions only in their hamlets while

Table 6.4
Access to Training by Wealth Category in the Two Communes

Groups	Better off	Middle	Poor
No. of households interviewed	10	19	34
No. of households attending trainings	7	19	22
Percentage	70	100	65

Source: CBUNRM survey, 2003.

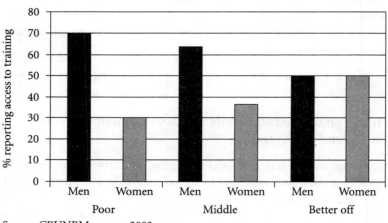

Figure 6.2
Access of Men and Women to Training Courses

Source: CBUNRM survey, 2003.

men attend courses in the commune centre or farther from home. Study visits are attended mainly by men. During the interviews, most women said they are responsible for housework, and visiting is for men. Local leaders and the organizers of study visits do little to change this situation; in the interviews, we learned that they believe that men learn better than women and that is the reason they are selected for study visits.

Our survey showed that only 30 per cent and 36 per cent of women in the poor and middle socioeconomic groups have access to training courses. Women in the better-off group have better access to training— equal to that of men in the same wealth group. In particular, better-off women have more opportunity to attend training organized by the women's union.

Traditional Training Methods

Farmers said that training is important to them because their production system is changing from shifting cultivation to settled farming. They are eager to attend training events, but stated that the knowledge they gain from conventional training is below their expectations. In one of our appraisal exercises with women and men farmers, we learned that farmers are interested in farmer-to-farmer exchanges and specialized training in production methods; they also said that they are able to learn and share experiences more rapidly and easily by working with researchers in the field.

The training courses and sessions organized for the villagers cover most crops of importance to them. On the other hand, conventional training sessions are held by agencies that are 'pushing' a limited number of commodities, including sugarcane, rubber and high-yielding varieties. These courses and sessions are developed by outsiders and designed to benefit government-initiated commodity-oriented projects. We found that most farmers did not understand the contents. Only about 34 per cent reported that they are able to apply the technologies that come with these commodities.

Conventional methods of training consist of lectures in a meeting room at the commune centre far from the fields or farmers' homes. The content of the training sessions is prepared mainly by extensionists and is based on knowledge gained from books or developed by researchers. Extensionists design the courses and sessions based on criteria from government programmes (Figure 6.3). Group discussions and the use of visual methods are very limited; only about 5 per cent of all training courses use videos or group discussion.

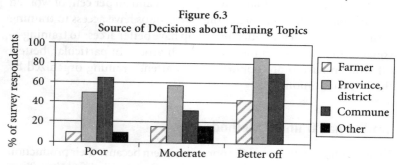

Figure 6.3
Source of Decisions about Training Topics

Source: CBUNRM survey, 2003.

According to officials and leaders, the content of most of the conventional training is suitable to local conditions. They argued that farmers should be able to apply what they learn in their production practices, for example, in rice and maize. However, the farmers interviewed said that the time devoted to training is too short; they said that the trainers are usually in a hurry to finish the courses and go back to the city. Combined with the barrier of language and their low education level, this makes it difficult for local people to understand what is being taught.

Perceptions of Local Leaders and Extensionists Regarding Social and Gender Issues

According to the local people and commune leaders, traditional training does not consider gender and social sensitivity in the process of technology development with farmers. All interviewed extensionists said that they developed activities based on the existing programme and contended that this addresses the needs of local people. Some extensionists said that they did a survey to learn about the problems of the local people. They also told us that they are aware that gender, age and economic situation influence the efficiency of training. However, in practice, almost all extensionists do their work based on the assumption that only those in better-off households are able to attend training courses or apply the technologies demonstrated during courses. Local people conform this. According to them, extensionists are never concerned about age, gender and economic conditions of people in the training courses and development activities. They often work only with better-off farmers to develop the technologies they prefer.

Although local leaders do not participate directly in training activities, they support extensionists at the local level by helping them select participants and by providing a venue (classroom). We held semi-structured and in-depth interviews with representatives of the Commune People's Committee, the women's association and the farmers' association. All interviewed leaders said that only commune and village officers and households who have a high level of education and production experience could attend training courses, because only these people are able to understand the course content. Some leaders said that it is better to invite men to participate in training and discuss activities with extensionists because they a have higher education level than women. They added that these men would train other farmers and also tell their wives about what they have learned. However, our gender analysis clearly shows that women often spend more time than men in agricultural activities. This reality is completely ignored in training sessions and related development interventions.

Overall, gender and social sensitivity and knowledge of both extensionists and local leaders are still limited. More appropriate capacity building seems warranted.

Building Human Resource and Social Capacity among Women and the Poor

We learned from our work that agricultural production is an integral part of CBUNRM or any rural livelihood system. Natural resources management in complex uplands production systems requires that both private and collective resources be managed in a complementary fashion. Building assets (which include access and rights to natural resource use) is essential to the process of poverty alleviation. Assets include a broad array of resources that enable people and communities to exert control over their lives, expand choices and participate in their societies in meaningful and effective ways. For example: financial assets such as credit and savings; natural resources such as forests, non-timber forest products, wildlife, land and livestock; social assets such as the capacity to build productive relations and organizations in the community and with the outside world (or terminate these links when they are harmful); and human assets, including knowledge and skills needed to gain access to services, markets, health care and other opportunities.

Upland people are often deprived of these assets. Being poor makes people less secure in terms of their livelihood and also reduces access to education opportunities, health services and other government programmes. Our project involved working with local people to understand their situation, support local organizations and build the assets of individuals in the community. In most communes social assets are constituted in formal and informal organizations, such as the Commune People's Committee, the farmers' association and the women's union. Our project supported and worked with all three organizations very closely as they have long-established relations with farmers and the poor. Many meetings were held to allow us to understand their roles, functions, strong and weak points, and to formulate development plans with them to help build and empower the organizations and their members.

Training in Gender and Social Issues

Training was ranked very high by both women and men farmers. Training in social and gender issues was provided to commune leaders, representatives of organizations and leaders of hamlets. Participants analysed the roles and activities of men and women in their community and in each household. Appropriate training was also organized for agricultural extensionists and researchers. Social and gender experts were invited to facilitate training.

Forming and Supporting Interest Groups

This work started with a diagnosis of local women's and men's needs and interests through group discussion and a brainstorming exercise. Farmers with similar interests were invited to form a small group to engage in a joint learning effort to address concrete needs or opportunities and improve their livelihoods. The following groups were formed by farmers:

1. **Rice production groups:** Many farmers, both women and men, are interested in wetland rice production to achieve food security. At the start of our work, this topic was their number-one priority. Two groups were established in the villages.
2. **Fish raising group:** Ten poor farmers who have a fish pond but lack capital for investment and want to improve management techniques formed a group.
3. **Pig raising group:** Women formed this group, which started with 20 members, but by the second year had grown to 48.
4. **Cassava and vegetable production group:** Many poor farmers are interested in these crops and formed a group to experiment with new varieties and improve management practices (including post-harvest handling).
5. **Home garden improvement group:** Seventeen farmers joined this group, which has become very popular and now has more than 60 members.

In the following section we report on our experience—with particular focus on the lessons learned—working with the rice and pig production groups. Rice continues to be the most important commodity for farmers in many parts of Viet Nam. Pigs are of particular interest to women farmers who see them as a way of earning a bit of extra cash for their other family needs. Extra cash allows them to pay for medicines or care when someone in the family is ill or perhaps to buy books or other school supplies. When people in a village are desperate, there is often a social system that they can rely on, but having extra cash is also extremely important for women.

Rice Production Groups. Each group initially consisted of no more than 15 farmers. Members began by discussing and explaining the problems they were facing and identifying possible solutions. The project team facilitated these discussions, using their scientific and technical knowledge to complement the farmers' local knowledge.

One of the problems identified was low rice yields. The group looked at the causes and discussed a range of solutions to be tested in their fields; for example, testing three new varieties (TH30, Khang Dan and D116), using the local variety as a control; applying various levels of fertilizer; and trying labour-saving transplanting and direct sowing methods.

Three to five farmers agreed to test one of the three options. Other group members participated in evaluation and learning meetings at least three times during the growing season: during planting and the experimental design stage; during the growth period; and at harvest time. At each meeting farmers developed their own criteria for monitoring and evaluating results and decided which varieties were performing well, how much and what type of fertilizer to apply, and which cultivation techniques to use. The results of on-farm monitoring and evaluation were shared with other rice farmers in the group as well non-members and other production groups. As such, the learning process was expanded to other farmers in the community.

Based on the lessons learned from these on-farm experiments with farmers, we believe that researchers should act as facilitators to help farmers develop solutions based on their own situation. Using our model, some farmers could then test new technologies while others monitor and evaluate the results and, thus, learn from those who are testing the options. The adaptability of options or solutions to more farmers should also be discussed by the farmers and supported by the researchers. The more familiar farmers become with new technologies and research results, the easier it will be for them to share lessons more broadly with others.

Pig Production Group. Due to lack of agricultural land, farmers wanted to increase their incomes by raising livestock. Some years ago a number of projects had supported and introduced cattle to the commune.

However, this only benefited middle and better-off farmers, who had sufficient money to buy better breeds or to pay herders to watch cattle. Poor farmers in Hong Ha and Huong Nguyen chose pig production as it is more suitable to their conditions. Pigs can be fed with farm products such as cassava, vegetables and other home-grown or collected feed. Some farmers had kept pigs in the past but their productivity was low. Pig production in upland conditions poses many problems, such as low performance of pigs, poor husbandry techniques, lack of suitable feeds and diseases.

Three experiments were carried out by different farmers: raising Mong Cai sows (a local breed favoured by farmers); raising cross-bred animals associated with the fattening of pigs for the market; and storing cassava roots and leaves as pig feed (cassava is widely grown). In addition, two farmers were trained in basic veterinary practices for one month in Hue. This was supplemented by follow-up training in the village with the help of researchers and students. Vaccinations were provided through veterinary service centres. Other group members visited the district market centres to gather information about the price of slaughtered pigs before they are sold to middlemen. The farmers decided to sell their pigs together to get a higher price.

Although these experiments were useful in building social capital, not all tests were successful. Mong Cai sows were provided to 10 farmers. Although the sows produced good piglets in the first year, after artificial insemination carried out by the university, subsequently insemination was not possible as there were no boars in the village. As a result, the farmers decided not to keep the sows. Instead, farmers now usually obtain piglets from nearby Huong Van, a lowland commune near Hong Ha where sows produce high-quality piglets. It is expected that some of the better-off farmers might begin to keep and breed sows and will then sell piglets to the others.

As members of an interest group, women and the poor have better access to credit from government banks or development projects. To build their capacity in credit management at the community and household level, we introduced a small revolving fund for each of the interest groups. Leaders and credit recorders in each group received training in credit management.

Coalition with Line Agencies

Cooperation between villagers and local organizations is important in enhancing villagers' management capacities (including their ability to negotiate). At meetings between farmers, women, extensionists and development programme officers organized in the communities during our project, villagers spoke out and requested support from the extension service and development organizations, and representatives of these groups informed villagers about the kinds of support they could offer. This type of encounter is instrumental in avoiding overlap or conflicts between organizations operating in a particular area. Farmers can also tell outsiders what they expect and how to organize activities more effectively. For example, a coalition between the extension service centre and the university could provide seeds and fertilizer to farmers (through the extension agency) and complementary technology training (through the university) at the same time. As another example: the province's animal and plant breeding enterprise could offer farmers fruit tree seedlings and the extension centre could organize a related study visit.

IMPACT OF HUMAN RESOURCE AND SOCIAL CAPACITY-BUILDING ON WOMEN AND THE POOR

What differences are these actions making? Based on our monitoring and evaluation efforts to date, our first conclusion is that the number of beneficiaries is increasing (Table 6.5).

Table 6.5
Number of Women and Poor Households Involved in Interest Groups

Interest group	Initial membership	Current membership
Pig raising	20	48
Rice production (2 groups)	15 in each	Almost all households
Fish raising	10	37
Home gardening	17	60
Cassava production	5	Almost all households

Source: CBUNRM fieldwork, 2003.

The number of farmers in each group has increased because the interest groups are providing an opportunity for farmers to learn from each other. For example, some farmers had never kept pigs or fish before, but since becoming a member of the interest group, they have attended training courses, learned from other farmers, and are applying this knowledge in their own production practices. We also found that the participation of women and members of poor households in technical training courses is now high. Before our interventions, the ratio of male to female participants was 1.5:1, but after a year of effort more women than men are attending technical training.

In the past only about 30 per cent of all participants in training sessions applied new technologies—mainly the better-off farmers. Our 2003 survey showed that this number had increased to an average of 43 per cent across the different wealth categories. However, 14 per cent still said that they were unable to introduce or test new technologies. According to the farmers, they appreciate learning about new technologies through the training sessions and field visits, but they often continue to experience difficulties due to language problems and a lack of experience in testing new methods.

Holding meetings near the hamlets increased attendance and promoted active engagement, in particular by women. Even when women were unable to attend such meetings, they would drop in as observers, then discuss what they heard and seen with their friends later. Researchers and extensionists should allow themselves enough time to stay in the community overnight and even participate in farm work in some cases. One of the team's female researchers was able to do this quite often and found it a very productive way to learn and build social relations and trust. In Box 6.1 we present some reflections made by women about their increased participation.

Local leaders have become more aware of and knowledgeable about social and gender issues. The establishment of interest groups has made them understand villagers' conditions better and the characteristics of different groups and farmers in their communes. They have also increased their capacities to support development activities (from planning to implementation and evaluation). More and more women and disadvantaged groups are involved in community development programmes, for example, in credit groups, as members of interest

> **Box 6.1 People's Voices**
>
> Since we set up a pig raising group, women have an opportunity to participate in training and learn from other farmers. Their production has improved steadily. (A leader of the women's union in Hong Ha commune)
>
> We do not have land for paddy and raising an income is difficult for us. But after we started participating the in the pig raising group, we have been able to keep pigs and ducks and get income from these activities. This compensates for the rice. (A woman farmer)
>
> I never kept pigs before, but when I learned about the possibility to using local feed stuff for pigs, I started raising them. Now I have obtained a very good income from pig raising and I am using this money to construct a new house. (A women farmer from Parinh hamlet)
>
> Training in rice production has increased the average yield from 2.0 ton/ha/crop (in 2000) to more than 4.0 ton/ha (in 2003). Many women and the poor families are happy with the training. They know how to improve rice production, raising pigs, and fish. (Vice-chairman of a commune in Huong Nguyen)
>
> Women attend the training sessions and study visits, and they apply the new knowledge in their production effectively. The income from women's activities is increasing. The position of women in the family is changing. Their husbands did not pay attention to housework before, but now they are sharing housework with their spouse more and more. (A representative of the women's union)

groups and in the agricultural land distribution process. The leader of one of the communes said:

> In the past, because of local customs, very few women and poor participated in technical training and other activities but now they are members of interest groups. This allows them to attend these activities.

Extensionists and development workers have tended to follow what their plans tell them to do. Now they have started to learn and practice a participatory approach. Extension agents are paying more attention to gender and social conditions of farmers in villages. The quality of their work is improving with more participation of women and the poor. They have begun to select venues and times that suit farmers better, especially women. They are changing their training methods from lecturing to visual and field practices by interest groups. Their new approach is based on what farmers need instead of what farmers lack.

To sum up, the number of women and the poor participating in almost all the activities has increased. Our gender and social analysis capacity and awareness building seem above all to have improved the *quantity* of participation. The most successful production activities are among farmers in the better-off group. Therefore, the question for the research team is how to improve the *quality* of participation of the women and poor. This will require more time and effort.

Fostering a New Approach: Conclusions

Commune leaders have told us that our approach is very different from other projects. Giving poor farmers, including women, the opportunity to improve their understanding and address *their* interests builds confidence and skills. In the past the ideas, priorities and local knowledge of commune leaders and other local people were mostly ignored by rural development 'experts' (researchers and extensionists alike). Our work suggests that poverty reduction in heterogeneous upland areas of Viet Nam is much more effective using participatory tools and fostering adaptive learning.

Our study shows that rural women and the poor have less access to information and services than men and better-off households in upland communities. The burden of agricultural production and housework, traditional norms and customs, as well as a lack of social and gender sensitivity on the part of extensionists, local leaders and researchers are important factors that have led to this situation. Changes will not come overnight. Improving their livelihood is the first priority of the upland poor. Not all farmers have the same interest and capacity to improve their production practices and income generation; so participatory approaches must make special efforts to engage all local people, especially the women and poor. Successful new technologies can best be disseminated by structured farmer-to-farmer learning activities and by extension agencies that use participatory tools and methods.

Improving the knowledge, awareness and perceptions of local people and leaders, researchers and development workers regarding social and gender issues through training, working together and sharing information contributes to a better approach to working with the poor and women. However, although we have seen a significant increase in the number of women and the poor involved in local development activities, this may still not mean that their voices are heard.

Improving the quality of participation of women and the poor remains a research challenge. It requires appropriate skills, experience and attitudes of (outside) facilitators to create the environment in which they can speak out. In addition to modifying our behaviour, we also need to continue to increase our willingness, awareness and knowledge, and improve our attitudes. More time and effort are needed, not only by the research team, but also by others—in government and non-government organizations alike.

Training in social and gender awareness, and knowledge for development workers and researchers is an important way to contribute to improving the situation of the rural poor and women. However, increasing awareness and knowledge is not so easy. It needs a sound monitoring and evaluation mechanism and process. It will also require institutionalization of the concepts and practices of participation and empowerment in rural development activities at all levels, from the grassroots to the national government.

Setting up interest groups of women and the poor, and enhancing their management capacity are good ways to build human resources and social capital in upland communities. Due to similarities in culture, customs and language among members of upland communities, the establishment of interest groups is a suitable way for farmers to learn together—not only about technical issues, but also about such aspects as management and market information. Involvement in these groups also increases farmers' confidence, especially the poor and women, through interactions in meetings and field visits. The quality of this learning process depends largely on the capacity of the group leaders to create an environment for sharing knowledge, information and experience. The women's interest group proved to be a good medium for interaction and learning among women. In this process, the role of the local women's union is important in maintaining the group; therefore, strengthening the capacity of the women's union is also important.

Finally, capacity building for the poor and women requires a longterm joint effort by all social actors involved in rural development—local authorities, governments, non-government organizations, researchers and international supporters.

REFERENCES

Food and Agricultural Organization (FAO) (1997). Gender: The Key to Sustainability and Food Security. Rome: FAO. Available at http://www.fao.org/sd/WPdirect/WPdoe001.htm (viewed 17 November 2004).

Hoang Thi Sen (2000). Gender Roles in Agriculture and Forestry Production in Hong Ha Commune. In Hue University of Agriculture and Forestry, *Community-based Upland Natural Resource Management Project Report 2000*. Hanoi: Agricultural Publishing House.

Luat, N.V. and D.K. Son (1992). Progress Report on Farming Systems Research in the Me Kong Delta. In International Rice Research Institute, 23rd Asian Rice Farming Systems Working Group meeting, International Rice Research Institute, Manila.

7

Herder Women Speak Out

Towards more equitable co-management of grasslands and other natural resources in Mongolia

Photo credit: Ronnie Vernooy (IDRC)

H. YKHANBAI, TS. ODGEREL,
E. BULGAN AND B. NARANCHIMEG

The Country and the Issues

Mongolia is a country with an extreme, continental climate. Located in Central Asia, it is bordered by Russia on the north and China on the east. With an area of approximately 1.56 million km² and a population of 2.4 million, it is one of the most sparsely populated countries in the world (0.6 person/km²). Grasslands, which make up about 82 per cent of the land area, are home to 24 million head of livestock (83 per cent goats and sheep, 17 per cent horses, cattle and camels) and 176,000 herder families (MNE 2002). Grasslands are the principal renewable natural resource in Mongolia, but they are fragile, highly susceptible to degradation and very slow to recover when degraded.

Historically and traditionally, private ownership of pasture land does not exist. Grasslands have always been state property and are used by herders or other groups according to their needs. During the Soviet era (1921–90) citizens had no right to own livestock. In exchange for a salary, they used state pasture lands to herd state animals based on seasonal grazing schedules and the pasture use regulations prescribed by collectives and state entities. In the early 1990s Mongolia began its transition from a centrally planned economy to a market economy, promoting democracy, decentralization and privatization. As a result of the privatization process, all livestock have been privately owned since 1992; grasslands, however, remain state property and are shared. Between 1992 and 2000 herder households increased 2.5-fold and livestock numbers increased by 17.5 per cent. At the same time, pasture management authority and responsibility devolved to the local level governments and herders.

According to some estimates, more than 76 per cent of the country's pasture lands are overgrazed and subject to desertification. The underlying forces causing these problems are considered to be climate change due to global warming; the increase in livestock after privatization (numbers continued to increase sharply until 1999); and the uncontrolled concentration of animals around water sources, settlement areas, hay lands and seasonal camps. Experience shows that herders like to increase their herd size and livestock numbers as a means of survival in a competitive market environment. Pastures and grasslands are a common resource and have low entry costs compared with other economic opportunities. After the breakdown of the state entities in 1992, this led many people to get into herding.

Problems

Overgrazing of grasslands and related natural resource management issues are serious issues that affect the ecological carrying capacity and management of all natural resources. The situation has become worrisome, both from a socioeconomic and an ecological point of view. The overuse of natural resources has a negative impact on the living conditions of herder families who depend on these resources. However, not all herder families are the same, and within families social dynamics vary (Ykhanbai et al. 2004).

Although both women and men play important, but different, roles in the management of natural resources in Mongolia's nomadic pastoralism, women's participation in natural resource use, decision making and implementation has been undervalued. In research and policy making, women's knowledge and abilities are often overlooked (Ministry of Social Welfare and Labour 2002). Only recently has women's participation in natural resource management been taken into consideration, and recognition of their work and key contributions is gradually increasing. Women are also slowly receiving more support for their activities.

TOWARD THE CO-MANAGEMENT OF GRASSLANDS

Co-management refers to shared decision making by (organized) herders and government. We believe that co-management of pasture and livestock herds is the best option in Mongolia. Pastures are used as a common resource and although private ownership of livestock is allowed (providing herders with the opportunity to develop a business-oriented livelihood), purely individual management is not in line with the carrying capacity of pastures. Individual management infringes on the interests of neighbouring herders who depend on the same resource base, and state ownership of grasslands requires government regulation of their use. The lack of capacity of herders and local government to manage the resource base sustainably requires participation and support of other stakeholders and sectors.

Several national policies and laws affect community-based natural resource management (CBNRM) and co-management of grassland resources. The new Land Law (2002) introduces long-term pasture use contracts for herder groups and communities, and the involvement of local governments. These contracts are only recognized if roles and responsibilities to ensure sound use, restoration and protection of the grasslands are defined *jointly* by herders and governors. In addition, in the Land Law and other legal documents, local governors are given the responsibility and the right to resolve conflicts concerning herders' movements between neighbouring districts (in Mongolia there about 330 districts or *sum*s) and provinces (*aimak*s).

The Sustainable Management of Common Natural Resources in Mongolia project—which is supported by IDRC and is being implemented by the Ministry of Nature and Environment (MNE) in collaboration with other ministries, agencies and NGOs—aims to improve herds, grasslands and natural resource management by implementing co-management approaches at selected study sites (Figure 7.1). The project is addressing this challenge using a combination of participatory and action-oriented field research in three of Mongolia's major ecosystems, representing varying degrees of population pressure and market access (Ykhanbai and SUMCNR Project Team 2004). Currently, more than 10 community or herder groups exist in the project area, with 13 to 32 herding families in each group.

Each group is considered a relatively homogeneous economic or social unit in terms of characteristics such as family, language, history, ecosystem and grassland area managed. At the selected study sites all stakeholders are participating in the co-management activities. Co-management agreements between the community and its members, and between the community and local governors are documented, with roles and responsibilities of all stakeholders agreed to during formal and informal meetings and discussions. A series of action research tools was used from the beginning, allowing individuals and others to understand one another better.

Sum-level co-management teams were established at each study site. They consist of representatives of herders, communities, local governors,

Figure 7.1
Location of the Study Sites

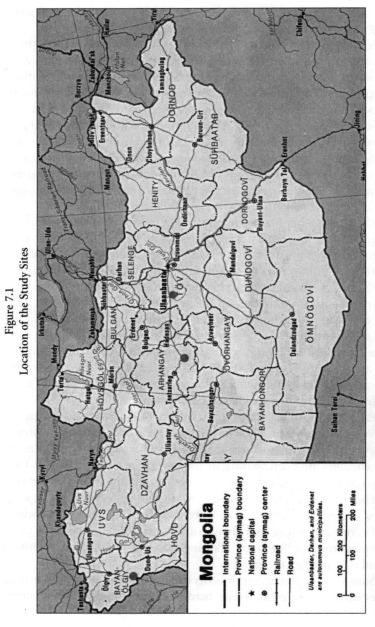

Source: Perry-Castañeda Library Map Collection.

NGOs, schools and other local organizations, as well as researchers from the project team.

In each community co-management contracts were established between the herders and the community (community contracts) and between the community and the subdistrict (*bag*) and district governors (herder–government contracts). These contracts can be revised anytime and are one of the elements of our adaptive management approach.

Herders are more likely to cooperate in pasture management when they also have an interest in working together on other activities that have a more immediate bearing on their livelihoods. Thus, the project team is also supporting and facilitating a number of other activities to improve livelihoods. These include: first, training in pasture management, natural resource management, vegetable growing, raw material processing, handicrafts, sewing, reforestation, seeding of hay lands, and participatory monitoring and evaluation. Second, the team encourages networking and experience sharing between groups, for example, through participation of herders in inter-site meetings and farmer-to-farmer visits. And third, the team provides institutional support, including small credit schemes, community funds, community projects, an information database in the *sum* centres, and supporting the set-up of community rules, community councils and women's groups.

Integrating Social and Gender Analysis

Our work follows a CBNRM approach, which implies the involvement of local people (both women and men) in all stages of the research work and cycle—analysis, planning, decision making, implementation, monitoring and evaluation. Social and gender analysis (SAGA) is used to identify and analyse the various roles, responsibilities and interests of women and men in pastoralism and other areas of natural resource management. In nomadic pastoralism women and men both play important roles in the management of natural resources. However, women's participation in natural resource use, decision making and implementation must be recognized more fully and their contributions valued more adequately. In many cases, women's knowledge and abilities are neglected.

Gender analysis allows one to identify and understand the different and changing roles, responsibilities and interests of women and men in pastoralism, agriculture, forestry and other natural resources domains (for example, hunting and gathering) (CBNRM Program Initiative 1998). Gender analysis highlights the human factor in natural resource management and shows that women's perceptions and interests are not fully included in decision-making processes at the family and community levels. Customs, tradition and religion play an important role in this regard. In Mongolia women assume many responsibilities for taking care of their families, tutoring the young, preparing food, making essential family goods, herding livestock and processing livestock products. Hence, their role and importance in promoting sustainable development for future generations are paramount.

A CBNRM approach is also concerned about action and fostering positive change. From a SAGA perspective, this means that women may need special attention in terms of support, training and other activities, and their decision-making roles need to be supported and strengthened. This leads us to an action-based agenda based on this understanding (McAllister and Vernooy 1999).

Our Agenda: Questions, Theory of Action and Methods

Our main research question was how to include the perceptions (ideas, interests and suggestions) of women's groups in more equitable co-management agreements for the use of pastures and other natural resources.

Our efforts were guided by the following key questions:

1. What does SAGA in natural resource management research mean for different stakeholders?
2. What key capacities are required by different stakeholders to allow SAGA?
3. What are the enabling and constraining factors?

The theory of action guiding our work included, first, to study the perceptions of stakeholders, including women's groups, and to identify the problems that women and men in the communities face. In particular, we aimed to pay attention to differences between women and men in perceptions, motivation and values regarding co-management. Based on the results of this step, we planned to support women's groups in the communities as a means to encourage economic and political empowerment. At the same time, we aimed to study how to implement effectively the concept of co-management and consider equity issues, focusing on women and men in decision-making activities and processes. We also planned to provide training to women to support their full involvement in the participatory action research. This study is one of the first research and development activities in Mongolia regarding the participation of women and men in natural resource management.

We expected the action research to result in:

1. Identification of women's and men's roles and responsibilities in the implementation of co-management and increased women's involvement in co-management teams at the *sum* level.
2. Assessment of women's and men's contributions to family income.
3. Consideration of the roles and responsibilities of women's groups in the whole process of contracting, implementation and monitoring of co-management agreements.
4. Identification of the enabling and constraining factors in herders' families in terms of various livelihood opportunities and development of alternatives for testing.
5. Exchange of experience by women's groups at the three study sites and learning from each other through cross-visits.
6. By the end of study period, establishment of co-management agreements between the community and the herders' families, the community and *bag* governor, the community and *sum* governor, and their revision according to the reviewed roles and responsibilities of all stakeholders.
7. Special production activities, mainly by women, to allow additional income benefits to women and their families.
8. Small revolving credit schemes managed by women (and by women and men together) with special attention to the poorer households.

We intended to involve and engage women's groups at the community level, other community members, *bag* and *sum* government officials, and NGOs. The role of the project team was to facilitate activities and carry out research in cooperation with the various stakeholders. Proposed activities in the community included: facilitating discussions among the herders' families; forming and supporting a women's group in each community for joint discussions about the inclusion of women's perceptions in the co-management agreements; and clarifying the roles and responsibilities of women and men in herder families. In addition, we planned to build capacity for taking responsibility in co-management agreement items in terms of implementation and monitoring of rules, rights and obligations, and organizing exchange visits between women's groups at the three study sites.

THE FIELDWORK

Study Sites

Mongolia with its vast land area, contains five main ecoregions: desert-steppe, steppe, mountain-steppe, steppe-forest and forest. Our project focused on the mountain-steppe, steppe and steppe-forest ecosystems in three *sums*.

Khotont *sum* in Arhangay *aimak* is in the Khangai steppe and steppe-forest ecosystem. This area enjoys strong neighbourhood support and social life, and the cultural, nomadic lifestyle tradition of the Khalha (the majority ethnic group in the country), women who are active in the co-management of natural resources, and a rich ecosystem, including taiga steppe and forest resources. The most urgent problem is that in some areas the number of animals exceeds the pasture's carrying capacity by 1.2 to 2.3 times in the spring and summer seasons. Continued overgrazing is not sustainable.

Deluin *sum* in Bayan-Ölgiy *aimak* is in the Mongolian Altai mountain and steppe ecosystem. Opportunities include the chance to build on the traditionally nomadic pastoralism culture of the Kazakh ethnic group, a strong pasture division system, and joint management of pasture and natural resources based on kinship relations. In 2002 livestock exceeded the capacity of the pastures in this area by 20 per cent, and this situation remains largely unchanged.

Lun *sum* in Töv *aimak* in the central Mongolian steppe ecosystem has good business opportunities and greater educational opportunities for women. Mixed ethnic groups represent incomers from other regions and ecosystems of the country. The most urgent problem in Lun is the very high density of animals owned by local herders, largely explained by its proximity to Ulaanbaatar, the national capital. The area is also visited by herders from other regions trekking their animals to the market in Ulaanbaatar. This double load leads to overgrazing of summer and spring pasture and to heavy desertification.

Currently, 10 communities are involved in the research, three or four at each study site, with an average of 20 herding families in each community. The communities are organized along a number of dimensions: socially (as neighbourhoods), economically (groups of herders) and ecologically (herder groups in one ecosystem). In 2001 the project started working with about 200 herders from 90 households in three communities. Today about 1,000 herders are involved in co-management efforts at the three study sites. Most households have four or five members (50 per cent); 26 per cent have fewer members and 24 per cent have more.

Methods

Participatory rural appraisal and action methods were complemented by a survey. Methods included focus groups, field interviews, oral histories, seasonal diagramming and field observations. Group discussions and field interviews were used to collect information on women's participation in the co-management of pastures and natural resources. Based on the results, co-management contracts among the main stakeholders were redesigned to include women's ideas and opinions. The gender analysis identified roles in natural resource management and decision-making processes at micro, mezzo and macro levels. The macro-level study also analysed government policies.

Inter-community exchange of the experiences of women's groups was facilitated and networking of women's groups was enabled with the support of the database centres at the study sites. A total of 461 women and men from 200 herding families in nine communities took part in the survey as well as community leaders, local governors, *sum*-level co-management team members and researchers.

Learning from the Field

Gender Roles in Natural Resource Management

Although both women and men play important, but different, roles in the management of natural resources in the country, women's contributions and participation in natural resource use, decision making and implementation have been undervalued. In many cases, in research and in policy making, women's knowledge and abilities are simply forgotten or neglected.

According to our survey, although women participate in animal husbandry, they are also occupied with many household tasks such as taking care of the family's food and clothing needs and ensuring the health of the children and other family members (Figure 7.2). This information corresponds with national data and the tradition of labour distribution in herding households.

Men usually do most of the work outside and away from the home, particularly selecting pastures, haymaking, herding animals, participating in meetings and business management. However, almost all of the men's work is seasonal. In contrast, women's work is continuous during the day and during the year. They usually do repetitive housework, particularly processing milk, taking care of children and housekeeping. In other words, the daily workload of women is higher than men, but almost all of this work is unpaid. Women's workload hinders their participation in community decision making and in meetings about natural resource management. From this we can deduce that by decreasing women's workload, it is possible to increase their participation in natural resource management.

Table 7.1 indicates that community members perceive that women play important roles in protecting natural resources, for example, by teaching their children about the environment and the traditional customs of appropriate use and protection of natural resources. The

Herder Women Speak Out: Mongolia 193

Figure 7.2
The Participation of Men and Women in Farming and Household Work (*n* = 84)

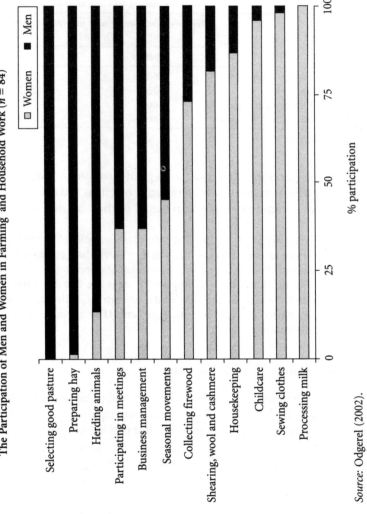

Source: Odgerel (2002).

Table 7.1
Participation of Men and Women in the Protection and
Restoration of Natural Resources According to Community Members

Task	Responses indicating who has responsibility (%) (n = 84)	
	Women	Men
Protecting the environment	68.3	31.7
Reforestation	73.3	26.7
Collecting forage plant seeds	100.0	0.0
Education of children in ecology	66.7	33.3

Source: Odgerel (2003).

common perception is that men have a greater role in decision making, for example, by participating in community meetings and making agreements with community leaders on behalf of their households.

> Mongolians have a long tradition of natural conservation. Women have great roles in keeping that tradition as they teach their children how to protect and soundly use nature. Also cleaning campsites, protecting rivers, and drinking and washing water areas are usually done by women. But the busy housework hinder women to participate in the participatory natural resource management. (Female member of a herder community)

Changes since Co-management

When co-management was first implemented, there was no common understanding of gender roles or equal participation by women and men in natural resource management. But this has now changed. Herders and other stakeholders have begun to recognize the importance of women's interests and their role in decisions about creating a sustainable system of co-management. At the same time, the knowledge of both women and men herders about the sound use and protection of resources and their skills and ability to do this have improved.

> Since the introduction of co-management, herders have become more likely to cooperate in natural resource management. Their knowledge and motivation to protect and restore natural resources have improved.

Especially women's roles and participation in natural resource management have increased in our *sum*'s communities. Now women and men equally participate in the adoption of co-management agreements. Women's participation in the decision-making and implementation process have increased; for example, the women group of Lun *sum*'s community initiated restoration of degraded winter and spring pastures through reseeding. (Male *sum* governor)

However, the division of labour between men and women has not changed, although women now participate more actively in community meetings and have begun to express their opinions more strongly.

Women's Groups

One of the project interventions has been the establishment or formalization of women's groups. These groups have increased women's opportunities to participate in decision making about natural resource management in their communities and are instrumental in identifying the best co-management practices and how obstacles to co-management can be overcome, building on local indigenous knowledge and practices.

Women's groups generally consist of all women members of the community. Their formation was facilitated by the project research team based on existing forms of collaboration between women. Each group has a leader, who is elected by the members at a community meeting. The groups meet once a month to plan activities, discuss problems and issues, and make decisions.

Some groups found it difficult to meet every month, especially those at the Deluin and Lun study sites, where some households are 50 to 100 km apart; these groups decided to meet once in two or three months instead.

> Women have clear roles in natural resource management. By establishing the women's group, women joined, and are sharing opinions, making joint decisions, and helping each other. (Female secretary of a community)

Women's groups have organized the following activities: supporting women's income-generating activities (for example, handicrafts, felt-making, vegetable growing), learning from each other (teaching their skills to other members of the community, learning from other communities, organizing various training for women in sustainable livelihood options and natural resource management), exchanging of experiences between communities and between study sites (community products exhibition, study tours to other sites, stakeholders' meeting about updating the co-management agreements in Khotont *sum*), and participatory monitoring and evaluation of the community's co-management efforts.

After the project team, together with specialists, provided pasture management training at the study sites, the communities designed a pasture rotation plan to improve their management system (see Table 7.2). Women's groups in Tsagaannuur and Arjargalant communities designed pasture shifting plans in agreement with the whole herder group, and these have been very effective so far.

> Women in our community made a pasture shifting plan. Women before never participated in pasture issues, but we saw that women's pasture plan was very clever. (42-year-old male community member)

As a result of the establishment of women's groups in all communities, they are more involved in community decision making; they

Table 7.2
Pasture Rotation Plan Designed by the Arjargalant Community Women's Group

What to do?	When?	Who organizes?
Shift from summer pasture to the banks of the Tsagaansum River	20 June–20 August	Community leaders, women's group
Some of the reserved summer pasture will be used for autumn	20 August–20 October	Community members
Otor movement to Ikh Nuur	20 December–20 February	Community leaders and members
Reseed winter pasture, planting steppe wheat grass and broom grass	Starting on 1 May	Women's group, community members

participate more in community meetings, freely express their ideas, and report about women's group activities to other members of the community.

Income Generation

The only source of herders' income is animal husbandry, which is under increasing stress as we described earlier. Creation of additional income-generating opportunities is very important as it can improve the herder households' livelihoods as well as contributing to solving the ecological problem of pasture degradation.

Location, proximity to market centres, infrastructure, and cultural and socioeconomic conditions at the study sites have a great impact on the implementation of co-management of natural resources. For example, in Deluin *sum*, culture, customs and close kinship relations became positive factors in implementing co-management. In Khotont *sum* herders' interest in finding alternative income sources is high because they have limited pasture area and small herds (herders also live close to each other) and the remoteness of a market centre requires cooperation. In Lun *sum* the households are more business oriented and interested in pursuing individual business interests because they are close to the market centre and have an established infrastructure.

The changing average income of herder households is illustrated in Table 7.3. In 2001, on average, 48.6 per cent of a herder's household income came from selling meat, 22.9 per cent from milk and dairy products, 15.2 per cent from wool and cashmere, 7.1 per cent from skins and hides, and the remaining 6.2 per cent from other sources such as pensions. Now herders are beginning to have new and additional sources of income, for example, from growing vegetables and making handicrafts. They have come to understand that it is not sustainable to have income only from animal husbandry. The series of recent severe winter (*zhud*s) that they had to face (2000, 2001, 2002), in which many herders lost their animals, have without doubt accelerated this realization. Comparing 2001 and 2003, herders now have new income sources including producing felt products (7.9 per cent of total income) and growing vegetables (4.1 per cent of the total income). These contributions may be increased during the coming years.

Table 7.3
Changes in Income Structure of Herder Households, 2001–03 ($n = 36$)

Source of income	% of income in 2001	% of income in 2003
Meat	48.6	41.5
Wool and cashmere	15.2	18.1
Milk and other dairy products	22.9	17.1
Felt products and handicrafts	–	7.9
Skins and hides	7.1	6.1
Other (e.g., pensions)	6.2	5.2
Vegetables	–	4.1

Source: Tserenbaljir (2004).

In herding households, women play the main role in generating additional income. The following interventions were facilitated by the project to assist in these endeavours.

Making Handicrafts with Animal Skins and Wool. Community representatives attended a training course in felt-making in the city and received wool processing equipment through the project. Community women make clothes and other handicrafts with felt for their own use or to sell. As a result, the incomes of herding households have increased. Women have also begun to have their 'own' money, which allows them to establish their own fund to support women's participation in natural resource management. Women have shown a strong interest in selling agricultural and other products, and their work has begun to provide visible and direct monetary benefits. In other words, their previously unpaid work is paying off!

Growing Vegetables. Communities at the three study sites are growing potatoes and other vegetables, which have become an alternative income source. These crops also help balance herders' diets. Vegetables are difficult to obtain in local markets, which partly explains why herders have taken to this experiment with such enthusiasm.

Small Funds to Support Women's Income-generating Activities. This year the project has funded women's small-scale income-generating projects and particular attention was paid to the poorer households.

Herders who live in the Khangai mountain forest-steppe ecosystem used to collect natural plants such as berries and medicinal herbs for their household consumption only. Now they protect these resources from illegal users and also gather them for marketing.

> I attended the felt-making training in Darkhan city. I learned to make good quality felt and felt handicrafts using the new equipment. I think that it is a very effective way to gain additional income. (Female herder)

> This year the community fund started to provide small loans for women's activities like making felt products and handicrafts and others. Thanks to this activity, households' income increased and women established their own fund. (Women's group leader)

The research team is currently looking into other options, such as establishing a rotating fund to support women's activities, to facilitate the marketing of dairy products and handicrafts, and to organize additional income-generating activities.

Women's Participation in Co-management

Community-based natural resource management can take the form of combinations of community-based and government interventions. Co-management is a governance arrangement that lies between a state property and a communal property regime. For co-management to be equitable, fair and equal participation of women and men in all activities is required. In many cases in research and in policy making, however, women's knowledge and abilities are neglected.

To clarify women's ideas about co-management, we surveyed the 461 women members of 220 herding families in nine communities. According to this survey, women define the following as important goals of co-management:

- to cooperate and agree on common goals;
- to plan activities and work toward certain goals;
- to improve knowledge and management of natural resources;
- to improve the use of pastures and other natural resources;
- to improve herding management and the productivity of animals;

- to improve and diversify the livelihoods and income of the households;
- to learn about the laws, agreements and rules related to herders and pastures; and
- to support and increase women's participation in co-management activities.

The surveyed women also proposed a number of ideas (ranging from 16 at one site to 36 at another) on how to include women's ideas and perceptions in co-management agreements. Women of Karatau and Buzaukol communities in Deluin *sum* defined their priority goal as 'to support and increase women's participation in co-management activities'. This was followed by 'to increase community income by processing animal raw products and making felt or handicrafts', and 'to increase community additional income by various agricultural activities like growing vegetables'. Fourth and fifth priorities were 'to increase women's knowledge about natural resource management by exchanging experiences with other communities or increasing newspaper availability' and 'natural resource restoration activities like improving degraded pasture condition by reseeding, increasing pasture water supply, and tree transplanting from dense parts of forest'.

Women of Arjargalant community in the forest-steppe ecosystem gave priority to 'restoring pasture, transplanting trees' and to 'decreasing pasture load by increasing income from other sources like growing vegetables', followed by 'activities for strengthening co-management'. In contrast, women of the Ikhbulag community (in the same forest-steppe ecosystem), which was established a year later than Arjargalant, said that 'increasing household income by making handicrafts by processing animal raw products' was their number-one priority.

Including Women's Ideas and Interests in Co-management Agreements

Although until recently women had fewer rights and opportunities to participate in collective decision making about natural resource management or other community affairs, most of them support the idea of co-management of natural resources if it is connected to improving their livelihood. Our work tries to increase women's role in this management aspect.

Early on in our work, some co-management agreements were drawn up without the roles of women being recognized. Subsequently, we sought to change these agreements based on the ideas of women. Generally, women's ideas on revising co-management agreements fall into two categories: representing their interests and improving the agreements.

Women's Ideas for Representing Their Interests in Agreements

- 'The community leader should include the ideas of men and women members in the decision making and he should share and introduce his ideas before making decisions.'
- 'Involve women in decisions about natural resource management issues.'
- 'Support women's participation in additional income-generating activities such as making felt and handicrafts, processing dairy products and growing vegetables.'

Women's Ideas for Improving Co-management Agreements

- 'Co-management activities among community and non-community herders should be organized and initiated by the *sum* governor.'
- 'The *bag* governor should meet community people once every quarter and convey their opinions and priorities to higher levels of government and also link herders' needs with local government policy.'
- 'Community leaders should take care of poor community members and take measures to increase the community members' livelihoods in relation to the sound use of the natural resources, their protection and restoration.'
- 'Community members who excel in sustainable management practices should be rewarded.'

- 'Clarify what measures should be taken if community members fail in their responsibilities.'
- 'Community members should cooperate, exchange experiences about natural resource management, and participate actively in community activities and training.'
- 'Marketing of community products should be organized jointly with the assistance of the community accountant and *sum* and *bag* governors.'

As women in the pastoral agriculture of Mongolia have traditionally played key roles in housekeeping and household work, many suggestions related to these tasks were mentioned but not reported here. This was expressed as 'to support women's participation in additional income-generation activities like making felt and handicrafts, processing dairy products, growing vegetables, etc.' or 'to highlight women's role in educating children and teaching them about natural resource management'. The most important ideas that women put forward were 'provide women with more possibility to participate in the community decision making and tell their ideas at the community meetings' and 'women's groups should be established'.

The women's groups listed three main positive influences of the co-management agreements related to pasture issues: the agreements are a good way for local people to implement the sound use of pastures and natural resources; good pasture management will make animals stronger and healthier; and the agreements increased options for the community to restore degraded pasture and natural resources, and to create watering points in the unused pastures.

In terms of negative factors, they mentioned four. They observed that there are often disputes with new or non-community people living in the area. It is also difficult to exclude people from moving to the community from other areas because of the harsh winter conditions. They noted that it is not possible to implement seasonal pasture rotation due to the shortage of pasture area (mentioned by the communities in the Hangai mountain forest-steppe ecosystem). Finally, they said that it is difficult to involve all community members in pasture protection given that some live far from each other.

Women identified some negative aspects of their participation in community decision making. When there is a community meeting, it takes a considerable amount of time to contact people and ensure full participation as some people live far away. Some decisions are not fully agreed on nor well implemented, as some members do not participate in the meeting and have no information about the decisions. Women are not unanimous on all matters: some said that 'community members actively participate and as a result now the community is successful', while others said, 'because of the lack of the active participation of members some community actions are not going well'. A few also noted that 'conflicts between members badly affect the community activities and the membership'.

Conclusions, Emerging Issues and Challenges

The most striking finding of our research to date is that stakeholders' understanding of the importance of gender equity in natural resource management has increased. At the same time, women's limited opportunities to participate in natural resource management and the domination of men's authority in decision making have been acknowledged. The main direct result of this study has been the revision and updating of grasslands co-management agreements between communities and local governors. These agreements now reflect community women's ideas and perceptions. They also recognize the need to fully involve women in future stakeholder meetings.

As a consequence of the establishment of the women's groups in the communities, women have more opportunities to cooperate, learn from each other, exchange experiences, and share knowledge and information. The establishment of women's groups has also facilitated gender equity in natural resource management and has created an environment to support women's participation in co-management of natural resources. It also encourages women's initiatives to protect natural resources according to traditionally inherited knowledge and customs.

Co-management agreements on pastures and natural resources are now being updated by including women's ideas and perceptions and promoting gender equity. As women have started to define their rights, roles and responsibilities in co-management agreements, they have become more actively and meaningfully involved in the community decision making.

Women's abilities in participatory monitoring and evaluation have improved as a result of the training provided by the project, and women's knowledge and abilities to generate additional income have improved as the result of the training and experience sharing.

Three important issues are emerging. Although significant progress has been made in women's involvement in the co-management of pasture and natural resources, men's authority still dominates. Hence, the assurance of gender equity in decision making remains a focal point. Second, although the traditional division of labour in pastoral agriculture has not changed much, since the transition to the market economy, women's agricultural labour burden has tended to increase. This hinders their full participation in co-management. Third, women's participation must be considered holistically and not only in terms of pasture management.

The main challenge is to continue awareness raising and transformative efforts. This includes continuing to support women's participation in co-management, further capacity building of women groups, and on-going exchanges of experience between the communities and among the study sites.

There is no doubt that Mongolian women have significant roles to play in sustainable development.

References

Community-based Natural Resource Management (CBNRM) Program Initiative (1998). Gender: Readings and Resources for CBNRM Researchers, Volume 1. Ottawa: International Development Research Centre.

McAllister, K. and R. Vernooy (1999). Action and Reflection: A Guide for Monitoring and Evaluating Participatory Research. Working paper, International Development Research Centre, Ottawa.

Ministry for Nature and the Environment (MNE) (2002). Sustainable Management of Common Natural Resources in Mongolia: Phase 2. Research proposal, Government of Mongolia, Ulaanbaatar.

Ministry of Social Welfare and Labour (2002). *National Programme on Gender Equality*. Ulaanbaatar: Government of Mongolia.

Odgerel, T. (2002). Report of the Survey on 'Labor distribution of Men and Women Herders in the 3 Study Sites'. Ulaanbaatar: MNE/Sustainable Management of Common Natural Resources in Mongolia Project.

Odgerel, T. (2003). Report of the Survey on 'Gender Roles in Natural Resource Management at Macro, Mezzo, Micro Levels.' Ulaanbaatar: MNE/Sustainable Management of Common Natural Resources in Mongolia Project.

Tserenbaljir, T. (2004). Report of the Survey on 'Community Herders' Income and Expenditures'. Ulaanbaatar: MNE/Sustainable Management of Common Natural Resources in Mongolia Project.

Ykhanbai, H., E. Bulgan, U. Beket, R. Vernooy and J. Graham (2004). Reversing Grassland Degradation and Improving Herders' Livelihoods in the Altai Mountains of Mongolia. *Mountain Research and Development*, 24(2): 96–100.

Ykhanbai, H. and the Sustainable Management of Common Natural Resources in Mongolia (SUMCNR) Project Team (2004). Sustainable Management of Common Natural Resources in Mongolia. Phase 2. Final report, Ministry for Nature and the Environment, Ulaanbaatar.

8

Similarities and Differences

From improved understanding to social transformations

Photo credit: Ronnie Vernooy (IDRC)

RONNIE VERNOOY AND LINXIU ZHANG

We conclude with two related chapters that look across the empirical findings of the six studies in search of commonalities and singularities. In this chapter we reflect on how social and gender relations inform natural resource management practices in the study sites and vice versa. We look at the main underlying forces or factors that shape particular management practices and some of the emerging issues in terms of equity and environmental sustainability.

One way to summarize the main results of the studies is along a continuum from descriptive to transformative, with reference to the key features of the 'women in development' (WID) and 'gender and development' (GAD) approaches mentioned in Chapter 1. This continuum goes from raising awareness about gender roles and gender-based inequity, to increasing knowledge about and skills in doing gender and social research and analysis, to the economic and political empowerment of women farmers and herders. Each of the six studies combines elements of this continuum, that is, features of both WID and GAD approaches. As such, on-the-ground realities seem to suggest that theoretical crossings are commonplace. We also identify a number of challenges that the studies have encountered. A long road to mainstreaming social and gender analysis (SAGA) in research and development and to more equitable relationships still lies ahead.

A quick look back at where we started from reminds us why it is important to integrate SAGA into natural resource management research:

1. To develop a better understanding and awareness of the social and power relations that govern access, use and control over natural resources. This involves understanding the differences and inequities among social actors and the particulars of local contexts.
2. To facilitate the recognition of the social and gendered nature of technologies, policies and interventions. Policies and technologies are value laden and influence women and men and different social groups differently.
3. To create room for social actors (women and men) to manoeuvre and to enhance the bargaining and negotiating power of groups that are marginalized and discriminated against, leading to empowerment and transformation where they have more access to, control over and benefits from natural resources.

Case Study Summaries

Before we start our search for commonalities and singularities, we briefly recapitulate the main characteristics of the processes studied and acted on in the case studies (Chapters 2 to 7).

Sikkim/West Bengal, India

Through its focus on ginger production and commercialization, the Sikkim/West Bengal study shows that social realities are often complex and sometimes contradictory. Gender dynamics interact with other social variables such as ethnicity and caste, and history is a major determinant of today's patterns. Cultural and ethnic identities and traditions (including taboos) inform the division of labour and mobility of women and men. The study argues that there is a high degree of socio-economic differentiation among households, and that among women new differences are emerging based on age and cultural changes. Across social groups, women have limited decision-making power and limited access to credit, which has a negative impact on their capacity to improve production. It is men who are entitled to land. Most women have only limited control over cash income and household expenditures.

The poor have limited access to land and, therefore, limited ability to expand production. In addition, the poor cannot afford to experiment (precisely because their access to land is constrained) and have difficulties obtaining government support. In Sikkim minorities (coming from Nepal) are often excluded from government support and research.

Nagaland, India

In the Nagaland case study we see an increasing demand for vegetables as a result of the rapid transformation of a subsistence economy to a market economy in combination with an urbanization process accompanied by changing food consumption habits.

Women are primary collectors of forest products and marketers of vegetables, but women marketers are differentiated economically and according to social status. Women vendors are facing constraints and hardships, partly because the market chain is not fully developed (for example, insufficient and inadequate transportation) and partly due to limited government capacities and poor service. Marketing is also a profoundly and often complex political-social process, in which many of the women vendors have only recently become entangled. The case study argues that, despite their significant roles and contributions to the economy, Nagaland women are not fully recognized or respected as economic agents and citizens.

Increasing demand is not without an impact on the natural resource base, which is being affected by the changes in society and the economy. Enclosure of vegetable fields and depletion of forest resources are leading to livelihood insecurity for some, while they represent opportunities for others—at least temporarily.

Nepal

In Nepal women play important but undervalued roles in farming, including seed production. The authors argue that the technology development process in Nepal, as directed and supported by the government, is largely gender blind and biased toward the rich. Only recently have some changes occurred at the conceptual level in terms of recognition of the relevance of SAGA. However, translating this recognition into practical steps to implement social and gender aware and informed policies and programmes remains a major challenge. At best, gender is defined as 'women', and 'women's' projects are seen as an adequate way to address gender issues. In research, those paying attention to gender are isolated and little integration takes place.

The field study shows that rural life is strongly shaped by social status and wealth hierarchies. Focusing on seeds, the study shows that the poor and uneducated have restricted access to both traditional and modern hybrid varieties, and this is having a negative impact on their livelihoods. Creating synergy between the informal and formal seed systems provides an alternative and, through project interventions, some progress is being made to increase access, improve incomes and maintain agricultural biodiversity.

Guangxi, China

The feminization of Chinese agriculture is one of the most significant features of rural life in the country today. At the same time, there is increasing disparity between the rich and poor (including in terms of access to vital services such as health and education). Ethnic minorities continue to be marginalized in many ways despite a number of supportive government policies. Their local knowledge and skills remain undervalued by society at large and by most researchers in particular. There is no system in place to recognize and value farmers' contribution to biodiversity conservation and crop improvement. The seed market, including that for local or locally improved varieties, is underdeveloped. Household food security remains unresolved and is becoming more problematic in places such as the uplands and in ethnic minority regions due to increasing globalization and free-market expansion. Environmental problems are widespread and serious. In terms of social and gender considerations, the study argues that there is a lack of awareness, understanding and planning by government, extension and research agencies, although some opportunities for change are opening up.

The Chinese research team is taking advantage of one such opportunity through a multi-stakeholder, collaborative effort. Action research interventions are focused on field-level experimentation with improved maize varieties combining (women) farmers' know-how with plant breeders' expertise. These experiments tackle productivity, environmental sustainability and empowerment. In parallel, action research aims to influence and change a number of key policies affecting rural livelihoods, including extension, seeds, *in situ* conservation, and intellectual property and farmers' rights.

Hue, Viet Nam

Strikingly similar to the Chinese and Nepalese studies, the case study argues that women in Viet Nam contribute more to agriculture than men, but their work is not recognized by the government. Most government policies and programmes are gender-blind, if not on paper certainly in practice. The study, with field data in hand, makes the point that, as a result, women and poor farmers have unequal access to education, services and information (for example, research results). The team argues that (upland) farmers do not learn much from current government services, such as extension, education and training, and that this negatively affects technology adoption and farmers' livelihoods.

The study documents and analyses how the design and delivery of rural development training are informed by gender and social variables. Many farmers (including women) say that men are better at learning about new farming technologies and practices. Local leaders and extension agents agree with this. They also think that men will disseminate knowledge automatically to women (one reason why they have no problem with only men attending training events). The study criticizes these gender-biased views and the outcomes they lead to.

Like China, Viet Nam is undergoing dramatic macroeconomic change. However, economic growth does not necessarily translate into reduced poverty, greater equity and environmental sustainability. On the contrary, gaps between cities and rural areas, between lowlands and uplands, and between the ethnic majority and ethnic minorities seem to be widening. Through participatory action research, the team is trying to do something about these problems. It supports local experimentation by interest groups and provides a different kind of training and other services to the rural poor and to women in particular. The team is also addressing a number of policy issues.

Mongolia

Mongolia is faced with serious degradation of natural resources, especially grasslands, making it increasingly difficult to maintain the traditional (but not backward) herding lifestyle. At the macro level, the country is undergoing rapid change (liberalization, urbanization, steps toward democratization), but the history of Soviet domination, the still heavily top-down government style, limits imposed by nature, and traditional practices and beliefs informing the gender division of labour and women's mobility constrain livelihood options and limit opportunities for change. Differences between the rich and the poor seem to be widening. As in the agricultural economies of Nepal, Viet Nam and China, (herder) women play important roles in natural resource management, but are undervalued by government and herders alike. Decision making is not shared equally, and power imbalances exist between women and men. There is widespread disrespect and disregard of local knowledge and practices.

The study team is pioneering a different approach to research and development, working toward social- and gender-informed policies and laws, participatory co-management agreements and livelihood alternatives. The team's efforts include a variety of field research activities, targeted training for herders, government staff and researchers in participatory rural appraisal, SAGA, participatory monitoring and evaluation, national and international networking and direct involvement in national policy making, including the drafting of new laws. Two innovative action research activities are the formation of community herder groups and the establishment of pasture co-management teams involving herders, local government and members of civil society.

Improved Understanding and Awareness

The key insights gained from the six studies include improved knowledge, the revaluation of local knowledge, and increased awareness about social and gender variables in natural resource management. The studies detail how rural households organize the production and marketing of a variety of crops (Nepal), vegetables (Nagaland), ginger (Sikkim/West Bengal), maize (China) and livestock (Mongolia). They document and analyse how gender roles shape the division of labour, as well as the processes through which income is generated and distributed or, sometimes, not distributed.

Gender roles were documented and analysed through the lens of

1. Men's and women's views on their roles in natural resource management, and their relations spanning the local and supralocal levels. As much as possible, study teams tried to listen to the 'voices' of women and men.
2. Methods and tools and how to use them in contextually and culturally appropriate ways, with an eye on historical changes.
3. Gender-based differences. Why do they exist? How and why do they change? In other words, how can one move from descriptive research and documentation to more explanatory research and writing?

4. Gender-based views and roles in the research process. This is an emerging and important topic. The studies presented here do not pay much attention to this issue yet, but the teams have expressed an interest in doing better in the future.

The Sikkim/West Bengal study highlights the changes over time and today's complexities in the interplay of ethnicity, gender, caste and age in the production and marketing of ginger. The study details how most women largely remain subjugated to men through particular notions of roles, duties and through rules of access to rural and urban spaces including fields and markets. However, some women, especially younger women, are creating opportunities for change.

In Mongolia men usually do most of the important work away from the home—selecting pasture, haymaking, herding animals, participating in meetings and business management. However, almost all of the men's work is seasonal. In contrast, women's work is continuous during the day and during the year. They usually do repetitive housework—processing milk, taking care of children and housekeeping. In other words, the daily workload of women is higher than men's, but most of their work is unpaid. Women's workloads hinder their participation in community decision making about natural resource management.

In all studies the differential impact of macro-societal forces, such as liberalization, privatization, commoditization and commercialization of service provision (health and education) shine through. These forces lead to many changes, such as out-migration of men in particular (China, Nepal) and related changes in the division of labour. Accordingly, workloads shift. Evidence suggests that, as a result, women tend to work more; and also do more unpaid labour. At the same time, some women manage to find new ways to increase income or at least explore such ways, for example, in Mongolia, China, Viet Nam and Nagaland. In all cases women continue to play a key role in unpaid processes of social production.

In Nagaland women are the primary gatherers of forest resources, such as varieties of wild vegetables, non-timber forest products, fuel wood and fodder, to meet household needs. The depletion of these resources, therefore, is having a direct impact on women, increasing their workloads and drudgery. It also has a direct impact on the overall livelihoods of the people as forest resources not only contribute to their life-support systems but also bring in crucial family income.

In China women and de facto women-headed households increasingly represent a disproportionate number of poor farmers. Women are becoming the main cultivators, food producers and on-farm income earners, but they also continue to play their traditional household roles. At the same time, women are underrepresented in government and administration, their significant roles are not recognized, and their specific needs, interests and expertise are largely neglected in agricultural and rural development policies and programmes. Their access to basic resources and services, such as credit, market information, training and extension services, remains limited.

A similar situation exists in Viet Nam. The Hue study reports that upland women spent about 10 times as much time as men on household chores. The amount of time women spend working in the field is also greater than men's. However, women have less opportunity to make decisions and less control over resources. There is a lack of awareness about social and gender factors in government development programmes. And men make most decisions about development-related activities.

In Sikkim/West Bengal significant changes are occurring due to the commercialization of ginger. Brahmin-Chhetri communities are gradually 'replacing' Rais and Lepchas in terms of forces of production. New land areas have been occupied and brought under ginger cultivation and new techniques have been introduced. Gender roles in cultivation, decision making and control over crops have changed as a result. Women's roles in the growing of ginger and control over its use have declined dramatically.

Although both women and men play important but different roles in the management of natural resources in Mongolia's nomadic pastoralism, women's particular roles and participation in natural resource use, decision making and implementation have been undervalued. In many cases, in research and in policy making, women's knowledge and abilities are neglected. Only recently has women's participation in natural resource management been taken into consideration. Women's roles in environmental management and preservation have taken centre stage, and recognition of their work and contributions is gradually increasing. Women are also slowly receiving more support in their work.

Several case studies highlight women's knowledge of plants and plant breeding, and their contribution to biodiversity conservation efforts:

From Improved Understanding to Social Transformations 217

the China and Nepal studies stand out, but we find elements of this as well in Sikkim/West Bengal and Nagaland. Here, the studies show how gender interfaces with class. In Nepal decision making about seed production and marketing shows that men dominate in richer households, whereas in medium and poor households decision making tends to be shared. Men play a dominant role in panicle selection in richer households; women are also involved among the medium and poorer households. Post-harvest operations like threshing and storage are exclusively women's activities irrespective of socioeconomic status. The seasonal migration of men from the middle and poor socioeconomic categories has brought about changes in gender role dynamics and resulted in greater involvement of women in seed selection.

The China study also pays attention to the knowledge and expertise of women extensionists and their courage to do things differently: from top-down, supply-driven service delivery to bottom-up, demand-driven service provision and very active roles in the research process. Some extension agents in Viet Nam are now also showing an interest in such an approach.

All six studies show how social relationships inform particular natural resource management practices, starting at the household level and extending up to the level of the village or beyond. A striking example is China, where male migration has become a prominent feature in many areas. As a result, speaking about farmers and farming means women-led and women-run farms. The studies also show how a number of women actively invest in new social relations to gain or increase access to resources; for example, the women vendors in Nagaland through their marketing interactions or the herder women in Mongolia through their interactions with the co-management committees.

In Nagaland, although at first most women are strangers in the marketplace, they make friends and learn from each other. Spending time in Kohima also helps women vendors gain knowledge about other town business; for example, they find out where to obtain the best prices for essential food items. In Viet Nam women have formed experimentation groups to try out new agricultural practices, receive training and technical assistance, and engage in marketing. In Mongolia women are joining co-management teams to have a greater say in the use and management of grasslands and other natural resources.

Although most studies have only touched on this area, we can hypothesize that the strong interest and active involvement of many women in the research process is another example of how social relations and opportunities are built through research. In China this has allowed access to new knowledge and seeds through contacts with extensionists, plant breeders and researchers from the Center for Chinese Agricultural Policy. The horizons are broadening.

For all the teams the research process itself has been instrumental in achieving results. Through face-to-face encounters with women and men farmers and herders at multiple sites (home, field, market and also events organized by the projects such as workshops, training sessions and visits), the researchers gained a better understanding and appreciation of women's and men's everyday lives and struggles, their knowledge, points of views, interests, worries and challenges. The fieldwork and the networking became roads to discovery and learning. The use of and experimentation with 'new' tools opened minds and eyes. Doing research *on* things and people shifted towards carrying out research *for* and *with* people. New research perspectives and practices are emerging.

The studies helped the researchers learn several major lessons: complex social and gender relations play a crucial role in farmers' adaptations and innovations. Unless this is understood and examined, the scientific and new methods introduced by the various institutions will not be successfully disseminated or accepted by the farmers. Integrating SAGA at the research level has to be the basis of any action research or development project (IDRC 2004d).

Other main stakeholders like plant breeders, extensionists, men farmers and, more importantly, policy makers involved in the projects have become more gender sensitive and now have a better understanding of women's roles in rural development and their specific needs and interests (IDRC 2004a).

TRANSFORMATIONS

Transformative agency refers to the ability to question, reinterpret, and change roles and responsibilities, and goes beyond doing things more efficiently; the latter could be called effective agency (Kabeer 2003: 174).

To varying degrees, the case studies have made steps toward transformative learning. This is an approach in which people together build a more integrated or inclusive perspective of the world. Through the learning process they jointly transform some part of their world-view (Van der Veen 2000; Vernooy and McDougall 2003). Manifestations of transformative learning in resource management include, for example, new values or patterns of decision making that farmers or herders generate and apply outside the immediate arena of the learning intervention. Transformative learning intentionally and consciously activates the praxis, that is, practice informed by theory, as a means toward (self-)empowerment of marginalized people and improvements in human systems.

The more transformative elements of the case studies include the formation of interest groups and their efforts toward the development of new or enhanced economic resources in the realms of production and marketing (socioeconomic empowerment); examples are the cases from China, Viet Nam and Mongolia. In addition, we note efforts toward more and more meaningful participation in systems and processes that allow the identification, prioritization and implementation of needs and opportunities, and mobilization around self-defined concerns and priorities (sociopolitical empowerment). We elaborate on these two forms of empowerment in the following sections.

New Economic Resources and Opportunities

Several of the teams are exploring new economic avenues to generate additional income as one way of reducing poverty. At the same time, these initiatives aim to strengthen (in particular) women's capacities as economic agents in their own right. In Nepal and China action research focuses on seed marketing; in Mongolia the focus is on adding value to traditional livestock-related products; in Viet Nam on producing and selling selected commodities, such as pigs and ducks. These activities do not have an economic focus alone. Through learning by doing, the women and men taking part in them are developing new or strengthening existing social and political skills, such as planning, budgeting, monitoring, negotiating, facilitating and communicating.

The Chinese SAGA action research team has supported the empowerment of a number of women farmers and strengthened their groups. These women have now become much more active in participating in and making decisions about major social and economic activities at the household and community levels. Women's involvement in open-pollinated variety seed production and marketing has built their confidence and capabilities; they have become more active and creative, and have taken the initiative to develop new income-generating and other activities in their households and communities as individuals and as a group (IDRC 2004a).

Building farmers' capacity in seed selection and quality control should be the first step to ensure access to quality seeds. The capacity building of community-based organizations in quality control at the field level ensures the sustainability of the seed production programme. Similarly, opening up the market for quality seed of local landraces provides incentives to the farmers and contributes to the conservation of local biodiversity (IDRC 2004c).

Collective Action and Empowerment

Individual forms of empowerment are critical, but in this book we have also seen the power of collective action to promote social change and gender equality. As Kabeer (2003: 194) has pointed out, this is an important route through which small-scale changes can be institutionalized at higher levels. Sometimes these collective actions take place in the political sphere: herder women entering co-management discussions and negotiations in the Mongolian study are a good example. More often, they do not take place in political arenas, although they are political in nature; the women's groups in Viet Nam and China are good examples. Women organize themselves to do things differently and challenge existing patterns and practices that inform production, reproduction and marketing. The Nagaland women vendors enter into politics through their daily struggles to sell vegetables.

Multi-stakeholder collaboration has a key role to play, but that is easier said than done. During the Second International Workshop in

Ulaanbaatar (see Chapter 1), this issue was synthesized in the form of the following chain of steps:

1. involvement of stakeholders;
2. detailed stakeholder analysis (define roles and responsibilities; set common goals);
3. voluntary participation;
4. development of strong common interests and ownership;
5. 'signing' of formal agreements (for example, the co-management agreement in Mongolia);
6. continuous review and adaptation; and
7. building of broader partnerships and networking.

To determine whether collaboration is moving in the right direction, it is critical to document the process systematically. Two of the key questions to address are: Who is (actually) collaborating? And how? Overall, the six teams made good progress in terms of following these steps; however, in a number of cases, more attention should be paid to working with and including specific stakeholders.

For the Mongolian team, this was their first experience doing SAGA research in relation to natural resource management. The researchers benefited considerably from hearing about the experience of others. Community members and local governors became more gender sensitive and their understanding about women's roles in natural resource management changed. After the SAGA work, women's groups became initiators of community-based natural resource management activities that aimed to increase women's participation in the decision-making process as well as in the co-management agreements. The co-management agreements were adapted after reflection on women's roles and responsibilities. Women's involvement in such activities increased. They have become creative and initiated income-generating and other activities in the pastoral households (IDRC 2004b).

The Chinese team reports:

> The women groups in our case study have also shown that farmers' self-organisation/management and autonomous capability are important.

They empower themselves in the mainstreaming process. In the context of China, strengthening women groups, or other farmer groups, and local communities can be the first step for such self-organisation and autonomous capacity-building. The collaboration between women groups and grassroots extensionists is an alternative to enhance women farmers' access to more institutional support and their linkages to formal (research, and seed) system. (IDRC 2004a)

SYNTHESIS

From the point of view of the women and men farmers and herders whom we have met in the previous chapters, SAGA means, above all, asking two questions: How do we manage to make a (decent) living while taking care of the land, water, forests and animals? How do we succeed in taking (meaningful) part in the socioeconomic and sociopolitical life of our communities? From these fundamental questions, others follow easily: How can we best use and manage our natural resources? How can we generate an income that allows us to feed and clothe our families, educate our children, and perhaps save or invest a small sum of money? How can we best organize ourselves to do this? What sort of tasks and responsibilities do we need to take on?

From a more analytical point of view, SAGA in the case studies has meant enquiring about the sexual division of labour and how notions of gender, class, caste, ethnicity and age inform this. It has also been expressed through attempts to carry out institutional natural resource management analysis, that is, identifying rules, rights, tasks, authority and control, claims on produce and income, and to look critically at policy-making processes that influence rural livelihoods and natural resource management in particular. And it has been shaped through participatory action research initiatives, such as the formation of experimental and interest groups, the building of new, multi-stakeholder alliances, capacity building and empowerment strategies, and informing or influencing policies through various means, including advocacy.

There are a number of striking commonalities. First, there is increasing pressure on the natural resource base due to multiple factors that include market liberalization, privatization and enclosure of the commons, state reform and structural adjustment, population increases and population shifts (urbanization). Although aspects of these factors differ among the six cases, there seems to be a common trend towards a more competitive environment, imposing restrictions on access and the generation and sharing of benefits. Several of the studies are trying to counteract this tendency and create more space for the disadvantaged. An impoverished resource base in its turn forces adaptations in the social fabric underlying production.

Second, the forces mentioned earlier seem to accompany or lead to increased social differentiation. This takes various forms: a wider gap between the rich and the poor, between those in urban areas and those in rural areas, between ethnic majority and minority groups; and feminization of agriculture (most evident in China). These widening gaps seem to be leading to more pressure on land, water and other resources: by the poor, because their situation tends to become more desperate; by the rich, because their demands become greater.

Third, ethnicity as expressed through ways of life, values, rules and norms emerges as a social force of relative importance. As all case studies indicate, ethnicity influences many aspects of natural resource management, but is often underestimated or neglected in research and policy making alike.

Fourth, continued and strong top-down policy-making processes affect local realities, very often making things more difficult for farmers and herders. At the same time, there seems some support for the search for (policy) alternatives and the opening up of opportunities for building new alliances between government and civil society.

References

International Development Research Centre (IDRC) (2004a). China Case Study Final Report. Ottawa: IDRC.
——— (2004b). Mongolia Case Study Final Report. Ottawa: IDRC.
——— (2004c). Nepal Case Study Final Report. Ottawa: IDRC.
——— (2004d). Sikkim/West Bengal Case Study Final Report. Ottawa: IDRC.
Kabeer, N. (2003). *Gender Mainstreaming in Poverty Eradication and the Millennium Development Goals: A Handbook for Policymakers and Other Stakeholders*. London: Commonwealth Secretariat; Ottawa: International Development Research Centre; and Hull: Canadian International Development Agency.
Van der Veen, R. (2000). Learning Natural Resource Management. In M. Loevinsohn, J. Berdegué and I. Guijt, eds, *Deepening the Basis of Rural Resource Management*,

Proceedings of a Workshop, 16–18 February 2000, pp. 15–22. The Hague: International Service for National Agricultural Research; Santiago: RIMISP.

Vernooy, R. and C. McDougall (2003). Principles for Good Practice in Participatory Research: Reflecting on Lessons from the Field. In B. Pound, S. Snapp, C. McDougall and A. Braun, eds, *Managing Natural Resources for Sustainable Livelihoods: Uniting Science and Participation*, pp. 113–41. London: Earthscan; Ottawa: International Development Research Centre.

9

Social and Gender Analysis is Essential, not Optional

Enhanced capacities and remaining challenges

Photo credit: Ronnie Vernooy (IDRC)

RONNIE VERNOOY AND LINXIU ZHANG

The six studies offer a rich panorama of ongoing fieldwork ranging from descriptive social and gender research to more transformative action research. The studies show some commonalities, but also considerable differences. These differences reflect how research context (history, geography, the knowledge and experience of the researchers, the knowledge and expertise of the farmers and herders), research questions and research methods interact in a particular way. They point to the importance of human agency, social differentiation and changing circumstances. They illustrate how gender interacts with other social variables, such as ethnicity, class and age. They provide evidence of enduring inequalities based on those social variables, but also offer encouraging examples of social change.

In this final chapter we reflect on the achievements made in terms of capacity building of both the researchers and others involved in the six research initiatives. We conclude with a brief discussion of remaining challenges. But before we begin, it is useful to quickly return once more to where we started: the many challenges involved in integrating SAGA into natural resource management including biodiversity research (as listed in Chapter 1) and the desire to do something about them through the Learning Studies initiative:

1. Knowledge of and experience in social science research among natural resource management researchers and research managers is limited.
2. Social science components are not well integrated with natural science components in most research efforts.
3. Researchers and research organizations have different starting points, interests and expertise in terms of social and gender issues.
4. 'Gender blindness' or the refusal to acknowledge the importance of gender issues is common in research and research policy making.
5. Short-term training has limited impact.
6. Resources in the area of SAGA and natural resource management in Asia are not widely available.
7. Networking has potential benefits but operationally is not easy.

Throughout this chapter, achievements and challenges can easily be linked to these seven points. As such, individually and together they serve as additional monitoring and evaluation reference points on our journey.

STRENGTHENING VARIOUS KINDS OF CAPACITIES

New ways of doing field research through participatory action methods and tools, farmer-to-farmer or herder-to-herder and other participatory forms of extension (for example, particularly in Viet Nam, China and Mongolia) and local development work stand out as achievements that can be located somewhere in the middle of the continuum from description to transformation. All teams experimented with methods that integrated a variety of tools, ranging from participatory rural appraisal tools to more qualitative research techniques. Some teams employed surveys (China, Viet Nam, Mongolia). The variety of methods served to address the different kinds of research questions teams were trying to answer.

All case studies contributed to increasing social science knowledge and experience among natural resource management researchers and research managers. The teams made great efforts to integrate social science components with natural science components. These integrative efforts also allowed them to triangulate information gathered. They also served the important function of allowing the teams to learn and become more skilled by doing. More practice will allow further capacity building and refinement of methods and tools to better suit local specificities.

The case study teams became engaged in a social learning process. Social learning is defined as the process by which multiple social actors with competing claims on a natural resource move towards, and engage in, concerted action at multiple scales of interaction (Röling 2002). It is about learning from each other. All six teams made considerable progress in terms of understanding the relevance of social learning and how to bring it about. The various facets of this achievement include the development and improvement, by researchers and others involved in the research, of negotiation skills (Nagaland), entrepreneurial and management expertise (Nepal, Nagaland, China, Viet Nam, Mongolia), facilitation skills (Mongolia, China), basic participatory monitoring and evaluation skills (Mongolia, Viet Nam, China), and advocacy skills

(Nagaland, Mongolia, China). Other skills that were instrumental for the teams include communication skills (all teams, but emphasized in the Sikkim/West Bengal team) and training skills (Viet Nam, Mongolia). All teams used and appreciated the networking skills they developed and improved throughout the research process.

The Viet Nam team reflected on its work as follows:

> We are using a variety of communication tools and participatory methods, including appraisal, monitoring and evaluation. Over time, we have seen the degree of farmer participation increase, and with a greater degree of farmer participation, results are more readily and widely adopted. We have also learned that farmers adopt simple techniques more quickly. As farmers gained confidence in the researchers, they accepted suggestions more readily and expressed their own ideas more openly.... Using a participatory approach has helped us direct attention to the role and needs of women. While men talk, it is often the women who do most of the work in the fields. Farmer participation is also useful in evaluating research results. Group meetings and on-farm workshops give farmers an opportunity for self-assessment. (From Chapter 6 of this book)

The Sikkim/West Bengal team reported:

> Support from the network and IDRC has been beneficial. The major support and learning from this has been in regard to the project proposal—the first workshop of the network [in Beijing, see Chapter 1] helped tremendously in focusing the proposal in a definite and clear manner. Second, the field visit by IDRC program officers was of great value and help by providing the researchers an opportunity to share their problems and difficulties. Discussions on the work and findings and the gaps and difficulties faced helped the researchers to resolve the problems and come up with the necessary solutions. Third, the sharing of work progress of the various teams of the network helped as well, by giving the researchers encouragement and ideas. Finally, the second workshop [in Ulaanbaatar, see Chapter 1] where all the teams and IDRC program officers met and presented the findings was invaluable in the sense that it gave the researchers a lot of input about how and what to do to improve their work. Not only this, it further provided us with an opportunity to share first-hand with the network the stories and experiences from the field. (IDRC 2004d)

Although the researchers learned much from the study, the women and men farmers, herders and vendors with whom they are engaged also strengthened certain abilities, such as marketing skills in the case of China, Nepal and Nagaland, and plant breeding skills in the case of China and Nepal. Or they acquired completely new ones, such as forming and managing an interest group or research group (China, Viet Nam, Mongolia). In some of the studies policy makers became more attentive to the needs and interests of (poor) women and men, such as in Nagaland, China, Viet Nam and Mongolia, and perhaps Nepal as well.

The Nagaland team wrote:

> Learning and capacity-building of stakeholders about the SAGA approach is at various levels. It is evident from the interactions with different stakeholders that at one point of time or another everyone has learned something. Some stakeholders were sensitized about the SAGA approach, and about the clamor for a negotiating space. Some are mainstreaming SAGA in the long-term planning process. By carrying out this SAGA action research, we, researchers, have enhanced our research skills and became more systematic in our approach to the research problems. We better understand the ground realities and have become more sensitive to the problems of the primary stakeholders, especially the part-time vendors who are vulnerable to exploitation and harassment. These part-time vendors are facing many challenges in order to get a marginal profit with their best farm produce. This research has also led to an emotional attachment between women vendors and us, strengthening rapport building with them. We have increased our negotiating skills to influence decision makers. We started channeling the 'voices' of vendors to the public using mass media and by interacting with policy decision-makers. (IDRC 2004b)

Networking has been useful in many ways. What the experiences highlight is the importance of learning by doing: implementing and adapting practices in the field with ongoing mentoring and support to address challenges and the continuous linking of experiences to conceptual (participatory) frameworks. The experiences suggest that strengthening the processes for peer networking, review and support, subregionally and regionally, toward the development of a 'community of practice' is a powerful means to build capacities. But learning by doing and the construction of a 'community of practice' are demanding in terms of both time and resources.

CHALLENGES

So far we have summarized the main capacity building achievements of the studies. Good progress is being made, but more learning lies ahead. In the final part of this chapter we highlight three challenges that have emerged.

Organizational Change and Mainstreaming of Social and Gender Research

Continuous awareness raising efforts, dynamic communications, incentives, active 'champions', examples of good practice and an effective monitoring system are some of the elements required to make sound social and gender research part of the everyday practice of a research organization. Although the six study teams are in one way or another working on one or more of these pieces, putting them all together still seems a long way off. It is clear that the teams and research organizations have different starting points, interests and expertise in social and gender issues; and within teams there sometimes exist important differences in terms of expertise, ideas and interests. Building bridges and working toward synergy even within the organizations and research teams remains a challenge, let alone doing so *across* organizations and teams.

All teams expressed a desire to do more work in their own organization and are looking for continuing support from IDRC and other donors. This is not surprising given that there are hardly any research organizations in the world that have a strong proven record. We let the case study teams speak for themselves. The China team firmly defines a principle:

> The case also illustrates that SAGA is essential, not optional, for the formulation of responsive and gender sensitive policies/regulations and related implementation and management to avoid further marginalization and biases in the mainstreaming process. (IDRC 2004a)

The Sikkim/West Bengal study team focuses on some key bottlenecks:

> Mainstreaming SAGA in the research world is crucial in order to get a holistic picture that will help in making/developing a purposeful development action plan/project to solve practical problems and to change undesirable and constraining situations. However, mainstreaming SAGA in the research world is a challenge as in almost all cases it is found that gender analysis is conducted only at the diagnostic level. And often it does not go beyond a description of the division of labor (roles of men and women). There is no analysis of the more important and complex gender and social relations. The reason for this is not only lack of capacity, but, more importantly, 'gender blindness' that exists in communities, organizations, institutions, individuals and even among the researchers. Many researchers do not recognize and take into consideration the differentiation of preferences, perceptions, responsibilities that exists on the basis of gender. It is either ignored or overlooked. This is one of the main challenges in mainstreaming SAGA in the research world. (IDRC 2004d)

The Nepal study team adds another perspective by looking at the world of policy making:

> Interactions with personnel from the policy level were not very easy. Getting an appointment was very difficult. There was a lot of confusion about concepts. The conversations repeatedly used to incline toward women's issues rather than about gender. Time and again, we had to bring our interviews back on track. This was not always appreciated by higher-level authorities. (IDRC 2004c)

The absence of a proper institutional framework or institutional body responsible for working on social and gender aspects was felt to be a major constraint in the research process. Some efforts of agricultural research and development institutions toward social and gender issues were scattered and lacked proper documentation. Meeting the right individuals and finding useful documents was time consuming.

There is another component of mainstreaming that merits attention, but one that is perhaps not so evident in the studies: communication of the study results to others involved in rural development, including extension agents, policy makers and implementers, and professionals working in education and training. The publication of this book is only one way to disseminate our research efforts and stories, the insights gained, and the achievements and remaining challenges identified. However, identifying other opportunities remains a task to be carried out in the coming period. Some of the teams are forming ideas already, including local language translations of this book (for example, China), and the use of chapters in teaching and training (for example, China, Viet Nam).

Enduring Inequities and the Empowerment of Women

Several cases document the enduring dominance of men in decision making, access and control, reinforced by conservative cultural norms and political systems, for example, Sikkim/West Bengal, Nepal, and also Viet Nam and Mongolia. In China we witnessed the feminization of agriculture, which is increasing the burdens on women, although at the same time the absence of men also is opening the door for some options at home. The six studies provide varied evidence of the interplay of various sorts of gender inequalities, in terms of access to resources, in basic human capabilities, in control of one's own labour, and, notably, in access to markets (all six cases).

The research efforts, with the possible exception of the Sikkim/West Bengal study, are also leading to some changes, although slowly and gradually. The key parameters for empowerment are access, control and a say in decision making. The cases represent a variety of empowerment strategies, such as organizing women's groups (Mongolia, China) or interest groups (Viet Nam) or both; capacity building, locally and via networking (Nepal, China, Viet Nam); building partnerships with stakeholders at other levels (Nepal, China, Mongolia); linking sustainable livelihoods with natural resource management, through attention to marketing and strengthening marketing links (Nagaland, Nepal, China, Mongolia); linking to policy making and policy makers (China, Mongolia); and bridging disciplines (all cases).

At the Ulaanbaatar workshop capacity building was further discussed, and the following steps were identified:

1. Identification of disadvantaged groups according to one or more social variables (class, caste, ethnicity, landholding, wealth, animal holding).

2. Identification of social motivators or animators (women and men).
3. Needs assessment: identifying not only needs but also confidence building opportunities.
4. Participatory action planning: who, what, how, where, when?
5. Implementation of an action plan.
6. Participatory monitoring and evaluation (indicators, local ownership, multi-stakeholder-based, intra-household relations and roles).
7. Participatory impact assessment: if desired results are not achieved, adapt the plan and process.
8. Documentation and dissemination: success stories; mistakes and failures; constraining and enabling factors, community 'voices'.

A key question is, whose empowerment strategy? It was noted that empowerment strategies, similar to research strategies, can vary considerably in terms of the nature of participation: from consultative to collaborative; from researcher driven to farmer/herder driven; from a focus on women (WID approach) to a focus on unequal relations in general (GAD approach). *Which* women and men participate (and which do not) remains a research question that requires more precise attention.

Empowerment is a work in progress. The cases vary in many ways, but there are also a number of similarities (for example, strengthening seed marketing, in particular by women, in the Nepal and China cases; see more examples given earlier). The 'whose empowerment?' and 'whose knowledge generation?' questions have been identified as important in all cases. The teams continue to explore how to move to a more action-oriented agenda, to counter socioeconomic and sociopolitical inequalities, and to reflect on the complex role of researchers as catalysts of social change.

Improving the Quality of Participation

Participatory action research can contribute to the creation of fora for analysis, discussion and negotiation in which ideas can be exchanged and initiatives planned. As the case studies indicate, this is why it is important to create opportunities for meaningful participation. The building of trust is essential, but this may take time and patience. The process of organizing often involves struggles over the definition of rules and norms, and researchers may become entangled in these struggles. Unfortunately, most of the case studies do not explicitly address this very important 'entanglement' or engagement question. We could hypothesize that what the Nagaland study points to is also relevant for the other studies, but this would require additional documentation and analysis. We highlight the last part of the quote used earlier in this chapter:

> This research has also led to an emotional attachment between women vendors and us, strengthening rapport building with them. We have increased our negotiating skills to influence decision makers. We started channeling the 'voices' of vendors to the public using mass media and by interacting with policy decision-makers. (IDRC 2004b)

This description suggests that participation itself remains a challenging research topic. It is not easy to stand back from the process and look critically at what is happening and how. Although most case studies made enormous progress in a relatively short time in terms of facilitating participation, there is likely scope for improving the quality of participation. As the Viet Nam team observed (Chapter 6), this requires appropriate skills, experience and attitudes of outside facilitators to create an enabling environment for people to speak out. Improving the quality of participation of women and marginalized people remains a challenge.

It is important to remember that information and knowledge are not value free and to be aware that the selective choice of information or knowledge may empower some people while displacing others. Knowledge is always socially constructed and often disputed. More analysis and reflection on the question of how different types of participation influence research results seem warranted. Mastering the art and science of participation is a life-long task.

The Power of Macro-forces

The studies provide various insights into how local communities are affected by processes at play at higher societal levels, such as commoditization, privatization, state reform and globalization. The China study refers to the WTO (the country's recent entry into this organization is an important issue) and asks if farmers, especially the poor, are favoured. The Viet Nam study notes that despite impressive and prolonged macroeconomic-level progress, poverty persists and overcoming poverty still remains a challenge in the development of the country. The study also highlights that poverty alleviation coupled with conservation of natural resources in the uplands is critical, not only for the local people but also for the nation as a whole. The same could be said about China—and Nepal.

The studies suggest that building linkages between local communities and national institutions and policy makers helps local actors demand more useful services and influence (to varying degrees) policy agendas. This includes the integration of government into the local planning process so that local interests and concerns are taken into account. It could also contribute to reorient the sourcing of technical assistance and the transfer of expertise.

More analysis will be required, but it is evident from the studies that these processes are not social or gender neutral. At the same time, the studies also demonstrate that these processes are not homogeneous by definition. Different responses occur and this is an important empirical finding that ought to inform theory development and refinement. Everyday differences, however small, matter!

References

International Development Research Centre (IDRC) (2004a). China Case Study Final Report. Ottawa: IDRC.
——— (2004b). Nagaland Case Study Final Report. Ottawa: IDRC.
——— (2004c). Nepal Case Study Final Report. Ottawa: IDRC.
——— (2004d). Sikkim/West Bengal Case Study Final Report. Ottawa: IDRC.
Röling, N. (2002). Beyond the Aggregation of Individual Preferences: Moving from Multiple to Distributed Cognition in Resource Dilemmas. In C. Leeuwis and R. Pyburn, eds, *Wheel-barrows Full of Frogs: Social Learning in Rural Resource Management*, pp. 25–47. Assen, The Netherlands: Koninklijke Van Gorcum.

About the Editor and Contributors

ABOUT THE EDITOR

Ronnie Vernooy is Senior Program Specialist, Environment and Natural Resources, at the International Development Research Centre (IDRC), Ottawa, Canada. He obtained his Ph.D. in the sociology of rural development from Wageningen University, the Netherlands, and joined IDRC in 1992. His research interests include rural development, natural resource management (including agricultural biodiversity), farmer and herder experimentation and organization, and participatory action research methods including social and gender analysis, monitoring and evaluation. Dr Vernooy has conducted and directed a number of rural development research projects in Nicaragua and currently contributes actively to community-based natural resource management research efforts in China, Cuba, Mongolia and Viet Nam.

Besides various articles, Ronnie Vernooy has authored, co-authored or co-edited several books. These include *Participatory Research and Development for Sustainable Agriculture and Natural Resource Management: A Sourcebook* (co-edited, 2005), *Seeds that Give: Participatory Plant Breeding* (2003), *Evaluating Capacity Development: Experiences from Research and Development Organizations around the World* (co-authored, 2003) and *Taking Care of What We Have: Participatory Natural Resource Management on the Caribbean Coast of Nicaragua* (co-authored, 2000).

ABOUT THE CONTRIBUTORS

E. Bulgan has a background in language studies. She holds an M.A. in linguistics from the University of Humanity in Mongolia. Since 2001 she has been working as secretary and research assistant for the Ministry for Nature and the Environment–International Development Research Centre project 'Sustainable Management of Common Natural Resources in Mongolia'. Her research interests include community-based pasture and natural resource management, participatory research, and social and gender research in natural resource management. She is hoping to pursue further studies in rural development in 2005.

Liz Fajber is a senior programme officer at the International Development Research Centre's South Asia Regional Office in New Delhi. She is active in programme areas primarily relating to rural development and natural resource management. Her interests focus on social and gender equity, access and tenure issues; local and indigenous knowledge and technologies; multi-stakeholder approaches; and enhancing community participation in, and benefits from, applied research. She has an M.A. in anthropology from McGill University.

Chanda Gurung holds M.Phil. and Ph.D. degrees from the School of International Studies, Jawaharlal Nehru University, India. The focus of her thesis was women's roles in development, women's rights and discrimination against them. Currently, she is based in Kathmandu, Nepal, and is actively involved as a researcher and consultant in various projects related to gender (natural resource management agriculture and indigenous knowledge); sustainable livelihoods; and participatory research and development. She has extensive experience working in the eastern Himalayan region (eastern Nepal and north-east India) as well as in the *terai* of Nepal. She has published several papers on gender and agriculture, and one concerning participatory approaches in agriculture. Gurung is also founding member and coordinator of the Eastern Himalayan Indigenous Women's Network, an NGO working on women's and gender issues in natural resource management and livelihoods.

Nawraj Gurung began his career as an extensionist with the Spices Board, Ministry of Commerce, Government of India. He now coordinates the horticulture programmes of the Swiss Development Cooperation/IC project in Sikkim. He is also coordinator of an NGO called Eastern Himalayan Initiatives. His professional interests include participatory technology development and methodological questions.

Hoang Thi Sen has a background in forestry and agriculture. She obtained an M.Sc. from Chiang Mai University, Thailand, and is currently pursuing Ph.D. studies in rural development at the Swedish University of Agricultural Sciences, Uppsala. She has been a trainer in participatory research since 1995 for project officers, development workers and farmers. Her current interests include farmer participatory research and gender analysis. She believes that social and gender research is not only about understanding a particular situation, but also, more importantly, about empowering local people and strengthening their capacities and assets.

Chozhule Kikhi is deputy director of training in the State Department of Horticulture, Nagaland, India. She is working with grassroots women on income-generating activities, such as home-scale food processing, organic vegetable cultivation and mushroom cultivation using agricultural wastes. From 1996 to 2000 she was the gender coordinator in the 'Nagaland Environment Protection and Economic Development (NEPED), Phase 1' project—the only female among 14 men. She has presented NEPED experiences and learning on gender issues nationally and internationally. She continues to work with farmers on the NEPED Phase 2 project to establish sustainable livelihood, especially for the people of the Angami and Zeliang tribes in Kohima and Peren districts. She holds a B.Sc. in home science from Punjab Agricultural University, Ludhiana, and has received considerable training in food processing and mushroom cultivation in Delhi, Bangalore and Solan, and in organic farming in Nishinasuno, Japan.

Le Van An is a senior lecturer at Hue University of Agriculture and Forestry (HUAF), Viet Nam. He is a livestock systems specialist and recently obtained his Ph.D. in animal science from the Swedish University of Agricultural Sciences, Uppsala (where he also did his M.Sc.). He is director of international relations at HUAF. His research focuses on rural development and livelihood issues. He has a special interest in doing participatory action research together with marginalized people. He coordinates a number of research projects funded by international donor agencies (including the Swedish International Development Agency, the Ford Foundation and the International Development Research Centre).

Vengota Nakro is deputy director in the State Department of Soil and Water Conservation, Nagaland, India. He is also a member of the 'Nagaland Empowerment of People through Economic Development' (NEPED) project's operations unit. The nature of his work is to collaborate with and support agricultural workers as they move towards a sustainable livelihood. His current assignments include activities with the people of the Konyak tribe. He holds a B.Sc. in agriculture from Punjab Agricultural University, Ludhiana, and an M.Sc. in tropical silviculture from Goettingen University, Germany.

B. Naranchimeg is a graduate of the Economics and Management School of the Mongolian State University. Currently, she is studying for an M.A. at the same school. She works as a researcher in the university's Population Training and Research Center with support from the Mongolian government and the United Nations Population Fund. She was an active member of the group on reproductive health research. Her areas of expertise and research interests lie in gender, development and demographic issues; poverty and environmental issues; and child labour.

Ts. Odgerel has a background in social sciences and holds an M.Phil. from the social science faculty of the Mongolian State University. She started her career as a researcher in the social and economic sector of the Research Center of the Mongolian Parliament. She has research experience in gender and social differentiation, Mongolian women's social conditions, women's labour and women's participation in natural resource management. Currently, she works as a researcher for the Gender Center for Sustainable Development, an NGO in Mongolia.

Pitamber Shrestha has a social science background. His expertise is rural development with particular interest in rural people's empowerment, local institution building and grassroots organization. He is also very interested in the field of participatory plant breeding, to which he has made contributions since 1992. Currently, he is LI-BIRD's site officer for the 'Strengthening the Scientific Basis of In-Situ Conservation of Agro-biodiversity' project in Nepal, a component of a global project coordinated by the International Plant Genetic Resources Institute,

Rome. This project aims to develop sound community-based biodiversity management practices for the sustainable conservation of agricultural biodiversity on-farm.

Deepa Singh is a horticulturist. Her research interests include agriculture biodiversity, natural resource management, participatory and action research and gender studies, and plant breeding. She focuses mainly on Nepal. She received her academic training at the University of Agricultural Sciences, Bangalore, and worked for LI-BIRD in 2003 and 2004. She is currently employed as a scientist by the Nepal Agriculture Research Council.

Yiching Song is a social scientist with a special interest in rural development, organization of (women) farmers and agricultural extension. She received her Ph.D. in communication and innovations studies from Wageningen University, the Netherlands. She also has extensive knowledge about participatory plant breeding and has been the project leader of a long-term research effort to create synergies between the seed systems of farmers and the Chinese government. Currently, she is a senior research scientist at the Center for Chinese Agricultural Policy in Beijing. She is the author of a number of journal articles and book chapters.

Anil Subedi is a rural extensionist. His research interests include rural development, natural resource management, agricultural biodiversity, informal seed systems, rural people's organizations and networking, participatory research methodology, appropriate technology, innovation and extension. He focuses mainly on Nepal and South Asia. He received his academic training at the University of Reading, United Kingdom. He was with the Lumle Agricultural Research Centre from 1971 to 1995 and executive director of LI-BIRD from 1996 to 2003. Since 2003 he has been the country director of the Intermediate Technology Development Group, Nepal.

H. Ykhanbai is director of the Forest/Pasture Policy and Coordination Department of the Ministry of Nature and Environment of Mongolia. He is the study team leader for the 'Sustainable Management of

Common Natural Resources in Mongolia' research project supported by the International Development Research Centre. A graduate of the Forest Engineering Academy in Saint Petersburg, Russia, he also holds a Ph.D. in natural resources economics from the academy. He has also attended environmental economics and macroeconomic policy natural resource management courses at Harvard University.

Linxiu Zhang is a senior research fellow and deputy director of the Center for Chinese Agricultural Policy of the Chinese Academy of Sciences. She has been working in the field of rural development policy research for more than 20 years. Her research focuses on land tenure rights and their impact on gender and resource management, rural labour market development, gender and poverty, and public investment in agricultural and rural areas. She has published widely in both English and Chinese journals.

Index

access and limitation, 59–60, 63
Agarwal, B., 22
agricultural research: social and gender perspectives in Nepal, 109–11
Agriculture Development and Conservation Society (ADCS) (Nepal), 123–25
agriculture extension work: social and gender perspectives in Nepal, 111–13
agriculture in China: commercialization of, 133–34; erosion of genetic diversity, 134–35; feminization of, 131–33; rural women and, 133–34
agrobiodiversity: conservation and use of, 101–26; management of, 105–6
aimak (province, Mongolia), 190
Arjargalant community, 196, 200; pasture rotation plan, 196

bag (subdistrict, Mongolia), 187, 189–90, 201
Bara (district, Nepal), conservation and use of agrobiodiversity in, 101–26
beez bhandar (Nepal), 117
Bhathi rice, 104, 113
Bhotias, 39, 47, 49, 62
Brahmin-Chhetri ethnic group, 41–45, 47, 54, 58–61, 63, 101; importance of ginger to, 47
Bulgan, E., 26, 34

capacity building: among women and poor in Viet Nam, 169–77
capacity-building programme, of IDRC: GAD approach in, 27, 29; objectives of, 25; project teams in, 26–27; questions agreed on for, 29–30; theory of action for, 27–28; WID approach in, 27
capacity building through SAGA: achievements, 228–31; challenges of, 231–37; in natural resource management, 227–36; power of macro forces and, 235–36; quality of participation in, need for, 234–35; strengthening of, 228–36
Cattle Trespass Act, Nagaland, 92, 96
Center for Chinese Agricultural Policy (CCAP), 26, 131–32, 136, 218
Central Pandam (village, Sikkim): ginger cultivation in, 52–55; size and ethnic composition of, 48–49
Chhetris, *see* Brahmin-Chhetri ethnic group
China: agricultural policy research context of, 135–36; commercialization of agriculture in, 133–34; feminization of agriculture in, 131–33, 143, 151; food security threat in, 134–36; genetic diversity erosion in, 134–35; local seed system strengthening in, 131–52; maize production in, 139; men and women's perceptions in, about decision making, 140–42; —, about management of households resources and activities, 140–42; needs and interests of women farmers, 142–43; research sites in, basic information, 137–39; rural women in, 133–34; women farmers' empowerment in, 131–52
Chinese Academy of Sciences (CAS), 26, 136

CIMMYT, see International Maize and Wheat Improvement Centre (CIMMYT)
Commune People's Committee, 169–70
Community-based natural resource management (CBNRM) programme: in India, 25, 30; in Mongolia, 185; —, changes since co-management, 194–95; —, gender roles, 192–94; —, in Deluin *sum*, 190, 195; —, in Khotont *sum*, 190, 196; —, in Lun *sum*, 191, 195; —, and income generation, 197–99; —, integrating SAGA, 187–204; —, learning from field, 192; —, pasture rotation plan, 196; —, research questions, theory of action and methods, 188–90; —, women's group, 195–97, 203; —, women's participation, 199–204
Community-based perspective approach, 159
Community-based Upland Natural Resources Management (CBUNRM) project, 157–60, 164; access to training courses, 166; conventional extension practices under, 165–67; perceptions of social and gender issues, 168–69, 178; rationale for integrating SAGA into, 159–79; traditional training methods in, 167–68, 178
Cornwall, A., 24

decision making, 140–42, 147, 159, 188–89, 192, 221; control and, 60–61
Deluin *sum* (Mongolia), 190, 195, 197, 200
Department of Agriculture (DOA), Nepal, 107
Department of Agriculture and Rural Development (DARD), Viet Nam, 162
District agricultural development offices (DADOs), Nepal, 111–12, 124–25

doi moi (reform policy, Viet Nam), 157

Eastern Himalayan Network (EHN), 26, 46

Fajber, L., 22, 25
farmers' seed system, see seed system
farmer-to-farmer exchange network, 106, 146, 148, 228
Farnworth, C.R., 23
food security, see genetic erosion
Ford Foundation, 136, 158

Gajo jotho vegetable, 73, 81
Gender and Development (GAD) approach, 27, 29, 35, 209, 234
gender blindness, 23, 227
Gender Research for Sustainable Development project (Mongolia), 26
genetic erosion: food security and, 134–35
ginger disease, 45, 49–55; management of, 56
ginger production and commercialization: access to and limitations of, 59–60; centrality of ginger, to people, 46–47; cultivation techniques in, 54–56; decision making and control in, 60–61; gender relations and ethnicity in, 59–63; —, cultivation practices, 45, 56–59; history of, at the research sites, 50–53; image and self-esteem and, 61–62; in Sikkim and Kalimpong, 37–63; inter-ethnic social relations and, 53–54; key issues and challenges in, 62–63; rituals observed for, 46–47, 51, 53, 61; roles of men and women in, 59; integrating SAGA into, 43–45; seed rhizomes in, 51–55, 57; site selection for, 47–49; social and gender nature of, 37–63; study objective and research questions on, 45–46

grasslands, 163; co-management of, in Mongolia, 183–204
Guangxi (China), 212, 214–18, 221–22, 231; maize germplasm collection in, 135; needs and interests of women and men in, 142–43; out-migration in, 132–33; seed fair in, 148–49; women's empowerment and local seed system in, 131–52
Guangxi Maize Research Institute, 145
Gurung, C., 26, 34
Gurung, N., 26, 34

haat bazaar: in Nepal, 108, 117, 125
herders, in Mongolia: co-management of grasslands by, 184–87, 194, 199–204, 222–23; fund to support income-generating activities of, 198–99; handicrafts of, 198; incomes of, 197–99; vegetable production by, 198
herder-to-herder extension, 228
Hoang Thi Sen, 26, 34
home gardens, 74, 90
Hong Ha communes (Viet Nam), 160–62; land use in, 162; population of, 162
Howard, P., 23
Hue (Viet Nam), 212–13, 217, 229
Hue University of Agriculture and Forestry (HUAF) project (Vietnam), 26, 158, 164
Huong Nguyen communes (Viet Nam), 160–62; land use and population in, 162
hybrid variety of seed, 110, 135, 139

In-situ crop conservation project, 103–4, 136, 152, 212
Integrated pest management (IPM) programme, 111–12
International Development Research Centre (IDRC) (Canada), 25, 28, 104, 136, 158, 231

International Maize and Wheat Improvement Centre (CIMMYT), 139, 145
International Plant Genetic Resources Institute (IPGRI), 103

jhum (shifting) cultivation, 66–67, 74, 82, 84, 90, 96
Jiggins, J., 23

Kabeer, N., 21, 24
Kachorwa village: crop diversity in, 103; geography of, 101–3; social and gender analysis of seed system of, 113–25
Kalimpong: ethnic groups in, 39, 41; ginger production and commercialization in, 37–63
Kharka-Sangsay village: ginger cultivation at, 50–51, 55–56; site selection of, 47; size and ethnic composition of, 47–49
Khonoma village, 94–95
Khotont *sum* (Mongolia), 190, 196–97
Kikhi, C., 26, 34
Kohima: gender roles in vegetable production in, 90–92; home delivery vendors in, 77; home gardens in, 74, 90; *jhum* (shifting cultivation) in, 74; Khonoma village in, 94–95; Merema village in, 90–95; Pholami village in, 79–90; terraced rice cultivation in, 68, 73–74; Tsiese Basa village in, 90–95; vegetable market in, 75–77; vegetable production methods in, 73–74, 90–92; vegetable retail shops in, 77; vegetable vending enterprises in, 75; wholesale dealers in, 76. *See also* vegetable vendors
Kohima Town Committee, 70, 72, 76–78, 88, 94, 96

Land Law (Mongolia), 185
landraces, 152; market system for, 122–24; of rice, 113–18

Le Van An, 26, 34
learning studies, project, 24–30, 227; road map for, 30–33
Lepcha ethnic group, 39, 41, 51–52, 54, 56, 58–62, 216; importance of ginger to, 46–47
Limbus, 54
livestock management, 140–42, 188
Local Initiatives for Biodiversity, Research and Development (LI-BIRD) (Nepal), 26, 103, 107, 110–11, 123
local seed system strengthening: in China, 131–52; formal and farmers' systems, 116, 126, 139–43; in Nepal, 101–26; SAGA integration to, 143–52. *See also* seed system
Lun *sum* (Mongolia), 191, 195, 197

maize, 134, 145; genetic base of, 152; germplasm collection of, 135, 139; hybrid varieties of, 135, 139; production of, 139
marketing capacity of farmers: in Nepal, 101–26
market linkages for women vegetable vendors: of Kohima, 66–98
mau rhizomes extraction and sale, 55, 57
Merema village (Kohima, India): vegetable production and gender roles, 90–92; women vegetable vendors in, 71–72, 78; —, enabling factors and constraints, 92; —, intervention options, 92–93, 96–97; —, part-time, 91
Mexico 1 (maize variety), 145, 148
Ministry of Agriculture and Cooperatives (MOAC) (Nepal), 107, 112
Ministry of Nature and Environment (MNE) project (Mongolia), 26, 185
Ministry of Social Welfare and Labour (Mongolia), 184

Mongolia: co-management of grasslands and natural resources in, 183–204; gender roles in natural resource management in, 183–204, 213–15, 217; men and women's participation in farming and household work, 192–93; men and women's participation in protection and restoration of natural resources, 192, 194; women's participation in natural resource management, 199–204; Land Law (2002), 185. *See also* Community-based natural resource management (CBNRM) programme

Nagaland: geographic and demographical issue, 67–69; natural resource management in, 210–11, 214–15, 217, 230
Nagaland Empowerment of People through Economic Development project (NEPED), 26, 70–71
Nakro, V., 26, 34
nangrey ginger, 50–51
Naranchimeg, B., 27, 34
Nasey village (India): ginger cultivation in, 51–52, 54; size and ethnic composition, 48–49
National Rice Research Station (Nepal), 108
natural resource management: capacity building through SAGA for, 227–36; gender roles in, 214–15; issues, 104–6; SAGA integration into, 19–35, 105–6, 187, 209–22, 227–36; social nature of, 19–20; sustainable management and, 20; transformation and, 218–22; trends and problems in, in Asia, 20–21; understanding and awareness of, 214
naya ko puja ritual, 46, 54, 58

Nepal: farmers' marketing capacity, 101–26; geography of, 101–3; integrating SAGA into natural resource management in, 105–6; local seed system in, 101–26; natural resource management issues, 104–5, 211, 214–15, 217, 232; Nepal Agricultural Research Council (NARC), 103, 107, 109–11, 117, 124
Nepal Rice Research Programme (NRRP), 124–25
Nepalese Seed Board, 106
Netherlands Development Agency (NEDA), 103
networking, 230
nomadic pastoralism, 184, 187

Odgerel, Ts., 26, 34
open-pollinated seed varieties (OPVs), 131, 144–45, 148, 151

parma system, 54, 57
participation quality: integration of SAGA for, 234–35
participatory monitoring and evaluation (PM&E), 32
participatory plant breeding (PPB), 146–48, 151–52
participatory rural appraisal (PRA), 32, 50, 72, 165
participatory varietal selection (PVS) experiments, 147
pasture land, 183; livestock herds and, 184
Peshore village (India): ginger cultivation in, 51–52, 54; size and ethnic composition of, 48–49
Pfutsero village (India): vendors in, alternative marketing chain of, 97
Pholami village (Kohima, India) women vegetable vendors, 71, 78–90, 96–97; constraining factors of, 87; costs and benefits of trip to Kohima market for, 85–86; enabling factors of, 79–87; family and community support for, 82; income and other benefits from vending vegetables by, 83–84; intervention strategies for, 88–90; non-material benefits of selling at Kohima for, 87; part-time, 84–87; retail and wholesale prices obtained by, 86; substituting vegetables for rice by, 89–90; transportation facilities for, 83, 87; vegetable production and, 80–82; vegetable species sold by, 80–81

Rai ethnic group, 41, 44, 47, 50–51, 54, 56, 58–63, 216; importance of ginger to, 46
Regional Seed Laboratory (Nepal), 108
rice: landraces, 113–18; production of, 113–14, 120–22
rural extension services, 152
rural livelihood system, 169
rural women, in China: commercialization of agriculture and, 133–34

sedentary farming, 163
seed fairs, 146, 148–50
seed-for-grain exchange, 118
seed-for-seed exchange, 118
seed system: capacity building and strengthening market links of, 122–25; exchanges of seeds in, 118–19; formal, 116, 126, 139–43; gender roles in, 120; informal, 116, 126, 139; knowledge and skills acquisition in, 122; market links of, 124–25; marketing channels of, 117, 121–25; men and women's participation in, 120; modern and local varieties in, 113–15; seed production in, 120–21; seed selection methods of, 118–19; seed sources of,

116–17; women's labour use in, 120–21
self-esteem: and image, 61–63
Shrestha, P., 26, 34
Sikkim: access and limitations, 59–60; Brahmin-Chhetris of, 42–45, 47, 54, 58–61, 63, 101; cultivation techniques in, 54–56; decision making and control by women in, 60–61; ethnic groups in, 39, 41–47, 51–52, 54, 56, 58–63; gender relations and ethnicity in, 59–62; —, in cultivation practices, 56–58; ginger cultivation in, local history, 50–53; ginger production and commercialization in, 37–63; —, key issues and challenges in, 62–63; image and self-esteem of women in, 61–62; inter-ethnic social relations in, 53–54; Lepchas, 39, 41, 44, 46–47, 51–52, 54, 56, 58–62; natural resource management in, 210, 214–17, 229, 232; Rais, 42, 44, 47, 50–51, 54, 56, 58–63; region, 39–41
Singh, D., 26, 34
slash-and-burn-farming, 163
social and gender analysis (SAGA): integration challenges of, 21–24; features of case studies of, 30–31; in agriculture extension, 111–13; in agriculture research, 109–11; in CBNRM approach, 187–90; in CBUNRM research, 159–79; in ginger production and marketing, 43–45; market linkages for women vegetable vendors and, 66–98; in natural resource management, 21–24, 105–6, 209–23; commonalities in studies on, 222–23; common issues in, 33; in Guangxi (China), 212, 214–18; in Hue (Viet Nam), 212–13, 217; in Mongolia, 213–15, 217; in Nagaland (India), 210–11, 214–15, 217; in Nepal, 211, 214–15, 217; in Sikkim (India), 210, 214–17; social transformation and, 218–22; in strengthening local seed system, 113–26, 143–52; of seed system in Kachorwa (Nepal), 113–26; learning studies project for implementation of, 24–30; learning studies roadmap for, 30–33; methods, tools and skill used in case studies on, 30–32; procedure adopted to study, 30–33; projects for, 24–30; rationale for, 159–79
social and gender issues: local leaders and extensionists perceptions of, 168–69, 178
social and gender research: organizational change and mainstreaming of, 231–33
social learning process, 228
social transformation, 218–22; through collective action and empowerment, 220–22; through new economic resources and opportunities, 219–20
Song, Y., 26, 34
Subedi, A., 26, 34
sums (districts, Mongolia), 185, 189–91, 196–97, 200–01
sustainable development, 134–35, 152
Sustainable Use of Biodiversity (SUB) programme, 25

Tashiding village (India): ginger cultivation in, 51–52, 54; size and ethnic composition of, 48–49
terraced rice cultivation (TRC), 68, 73–74, 89, 96
Thakuris, 42–43
Tsagaannuur Community, 196
Tsiese Basa village (Kohima, India), 71–72, 78; gender roles in vegetable production in, 90–92; intervention

options for vendors in, 92–93, 96–97; part-time vendors in, 91

upland communities: development projects in, 157–67; —, traditional extension tools for, 165–67; perceptions of local leaders and extensionists regarding social and gender issues, 168–69, 178

vegetable production: gender roles in, 90–92; home gardens for, 74; income from vending and, 83–84; insect infestation in, 83; *jhum* cultivation and, 66–67, 74; methods of, 73–74, 90–92; species in, 80–81; terraced rice cultivation and, 68, 73–74, 89
vegetable vendors (in Kohima, India): alternative marketing chain of, 97; constraining factors for, 87–88, 92; enabling factors for, 79–87, 92; home delivery vendors among, 75, 77; intervention strategies for, 88–90, 92–95; in Khonoma village, 71, 78, 94–95; market linkages for, 66–98; market regulation of, 77–78; in Merema village, 71–72, 78, 90–95; Naga vendors, 75–76; part-time vendors among, 75–78, 84–87, 91, 96; in Pholami village, 71, 78–90; regular vendors among, 75–78; research on, objectives and purpose, 70–72; retail shops and, 75, 77; in Tsiese Basa village, 71–72, 78, 90–95; vegetable production and, 73–74, 90–92; wholesale dealers and, 75–76
Vernooy, R., 22, 25
Viet Nam: ethnic minorities in, 157–58; forest policies of, 158; human resource and social capacity building among women and poor in, 169–77; perceptions of local leaders and extensionists in, on social and gender issues, 168–69, 178; reform policy in, 157; wealth ranking of commune households in, 163
Virtual Resource Centre (VRC), 30

wastelands, 163
women: access to information and technology, 159; access to training courses, 166; access to inputs in cultivation, 59–60; empowering of, 131–52, 233–34; enduring inequities faced by, 233–34; impurity of, belief in, 45, 53, 60–61; participation in management of natural resources, 199–204; programmes for, 111; roles of, 59; self-esteem and image of, 61–62; social capital strengthening of, 157–79; as vegetable vendors in Kohima, 66–98; —, potential interventions for, 94–99. *See also* women and poor (in Viet Nam), women farmers (in China)
women and poor (in Viet Nam): access to information and services, 177; cassava and vegetable production groups among, 171; coalition with agencies for, 174; fish-raising groups among, 171; gender and social issues training of, 170; home garden improvement group among, 171; human resource and social capacity building among, 169–74; impact of human resource and social capacity-building on, 174–79; interest group formation and support for, 171–74, 178; pig-raising groups among, 171–73; rice production groups among, 171–72
women farmers (in China): clarification on development of new seed varieties, 151–52; involvement in OPV seed production and

marketing, 148, 151; linkage between women's group and extension system at grassroot level, 147–48, 150; participatory planning, 144–45; policy issues, 150–52; reorienting rural extension services, 152; SAGA of, 143–52; seed fairs for, 148–49; skill enhancement of, 146–47; sup-port to establish viable seed enter-prises, 151

Women for Development Division (WFDD) (Nepal), 107, 112
Women in Development (WID) approach, 27, 29, 35, 209, 234
World Trade Organization (WTO), 131, 133, 150–51

Ykhanbai, H., 26, 34

Zhang, L., 26, 34